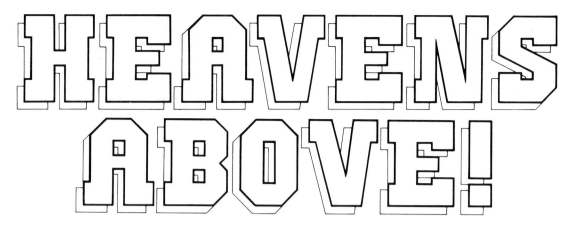

HEAVENS ABOVE!

Heather Couper and Terence Murtagh

A spectacular view of Saturn, its rings and satellites Tethys and Dione, photographed by Voyager 1 during its flyby in November 1980.

Franklin Watts
London   New York   Sydney   Toronto

Design and illustration copyright © Franklin Watts Ltd 1981
Text copyright © Trident International Television Enterprises Limited 1981

Franklin Watts Ltd
8 Cork Street
London W1

Designed by David Jefferis

ISBN UK edition 85166 909 3
ISBN US edition 531 04287 1
Library of Congress Catalog Card No: 80–54725

Phototypeset by Tradespools Ltd, Frome, Somerset
Printed in Great Britain by Sackville Press, Ltd, Billericay, Essex

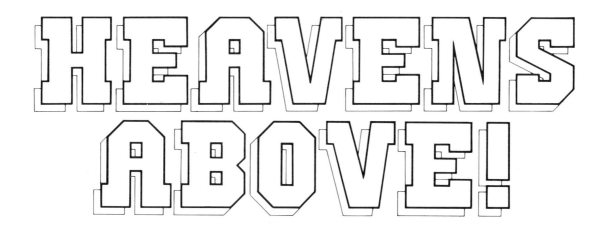

Contents

---**1**---

Voyage of the Starbreaker...4

---**2**---

The Solar System...12

---**3**---

The view from Earth...30

---**4**---

Stars...40

---**5**---

The scale of the Universe...46

---**6**---

How astronomers work...52

---**7**---

Is anyone there?...56

Watching the skies...60/Going further...62
Index...64/Acknowledgments...64

1: Voyage of the Starbreaker

You are about to join *Starbreaker* in a journey through space and time. Our mission will begin in a remote galaxy known by its inhabitants as OVOG-1: a galaxy so well explored that its more advanced civilizations feel the need to break away and discover worlds beyond.

Although our journey is imaginary, the places we visit along the way are real. And toward the end of the journey, you should be finding yourself in familiar territory.

Controller's Log. Project SEARCH. Starfleet Ship 1: STARBREAKER.
Interval 1

Project SEARCH begins.
Mission: to explore nearby galaxies beyond our home system, OVOG-1; to investigate their stars, planets, moons and lifeforms—if any.

Our fleet numbers six of the biggest relativistic starcruisers ever built. There are six destinations, all unexplored. Aboard *Starbreaker* we are bound for the huge spiral galaxy GC 8—Lacticos—which lies 2¼ million lightyears away. It is 470.00 Local Time ZD 325673 as we push away from Orbiting Station 5.

Interval 2
We're on our way, speeding ever more quickly through the familiar suburbs of OVOG-1. We pierce dark dust clouds like a laser beam and hurtle past cocooned young stars which look close enough to touch. We're accelerating so fast that the planet time we are used to will soon have no meaning for us. Our scientists have proved that if you travel close to c, the speed of light—300,000 km/s (186,000 miles per sec)—time itself will run slow. *Starbreaker* can make 99%c! And so, slicing through space at 295,000 km/s (183,000 mps), our journey of over 2 million lightyears will take only a fraction of our lifetimes.

Interval 3
The stars are beginning to thin out; we must be close to the edge of OVOG-1 now. Just the odd red dwarf star to report. Now and again there are dense globular clusters made up of thousands of incredibly old stars.

Interval 4
Cruising speed of 99%c achieved. OVOG-1 lies behind us now, and we have become the first intergalactic starcruiser. Swiftly past OVOG-2, our small elliptical companion galaxy. Ahead, there is nothing.

Interval 5
Nothing. Empty. Just space. No gas, no stars, no planets, no moons, no . . . anything. Between the galaxies there is a void.

Interval 6—500
Even at this speed the journey seems endless when you're in the void. But we can't go any faster: the speed of light is the speed limit of the Universe.

Interval 501
At last—the void is over! Today we picked up the spiral arms of Lacticos on our 21-cm-line monitor. It's growing bigger every day as we approach. We're preparing to start deceleration and enter the galaxy.

Interval 502
Steady deceleration is continuing, and time is starting to run at a normal pace. Lacticos and her two companion galaxies are completely filling the sky. It's so much like OVOG-1 that we're feeling quite homesick. But there's no time now for sadness. We must just make one final check on all systems, co-ordinates, detectors and guidance mechanisms, and then it's into an uncharted territory to the target of our mission where we expect to find ... what?

Into the galactic deeps

Interval 503

Now we're in the graveyard of Lacticos, moving steadily toward the ghostly spiral wheel of stars and gas hanging tantalizingly below us. It is vast. Lacticos is made up of a hundred thousand million stars arranged in a great spinning wheel so wide that even light would need 100,000 years to cross it. Up here we're surrounded by the grim remains of dead and dying stars, but there are compensations in the scores of globular clusters.

At the moment we're passing one which must have almost a million stars. What it must be like to live in the middle of a swarm like that!

Interval 504

Strange! When passing that globular, we were doing a routine check on our radio receivers and we picked up something very odd. It was a very faint signal, and we swore it couldn't have been natural. First it was pulsed: on-off, on-off. Then it was delivered in an ordered way, as if it was some kind of code; and the whole thing kept repeating, again and again. But it wasn't coming from the globular.

It came from the disc of Lacticos down below us. It was almost as if something down in Lacticos was

▼ Young stars: the Seven Sisters were born only 60 million years ago.

▼▼ Old stars: globular cluster M 13, made up of a million old red stars.

beaming up a message toward that globular! Sounds bizarre, but we're investigating. First we've got rough co-ordinates for the source, so we're heading in that direction even though the beam is so wide we can't be spot-on. Second—in case it is a message—we're trying to decode it.

Interval 505
We're headed on a trajectory toward Lacticos Co-ordinates 852048. It's a spot in the outer part of the galaxy, but now that we're down among the inner spiral arms, it's difficult to see a clear picture. There's plenty of activity here! On all sides we are surrounded by starbirth: billowing gas clouds concealing hot pockets of young stars; youngsters flaring fitfully, still learning to shine steadily; and jewel boxes of stars in cradle-like clusters, ranging from electric blue through gold and orange to dim red. Mixed in with the youngsters, we can pick out the odd member of the older generation. Here and there is a soot-choked red giant; and sometimes we see thin wisps of gas adrift from the wreck of an exploded star. But these exotic beasts won't help us in our search for planets—for life. Inhabited planets prefer to live only around coolish, unchangeable stars; very ordinary stars, in fact.

Interval 506
We're trying—without success—to crack that signal we picked up. It definitely *is* a message. Seems to be a picture of some kind....

Interval 507
A narrow escape! At 361.9 Interval Time, *Starbreaker* began to be pulled off course—but there was nothing there to pull it! Computer 7's trajectory simulator module gave us the answer in milliseconds: a black hole at LC 693050. Only a small one, so we were able to pull away in time, but when we looked back, what a sight! That pinprick black dot was inescapably locked in a dance of death with a vast blue supergiant star, whose surface layers were being dragged into a whirling vortex by the hole's enormous pull of gravity. What happens when the stuff goes down the hole is anyone's guess: it certainly can't get out again.

Interval 508
Starbreaker is homing in on LC 852048. Co-ordinates correspond to a yellow dwarf star, intermediate between spiral arms 3 and 4. We've done a preliminary spectral survey on it, and it looks good: temperature 5,800°C; age about 5,000 million years; normal size, mass, luminosity; no short-term light variation. LC 852048 is just the kind of star we'd expect to have inhabited planets.

◀ The intergalactic star-cruiser *Starbreaker* speeds towards the target of its mission, the galaxy Lacticos.
▼ Dead star: the Veil Nebula, ancient starwreck.

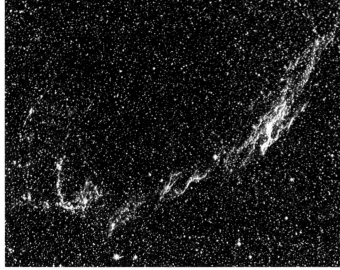

Nine worlds and a star called Sol

Interval 509
We're in the LC 852048 system now, and we're pretty certain there must be planets. There are enough comets! Today we had to steer through a huge cloud of frozen cometary nuclei at a distance of about one lightyear from the star, which augurs well for solid stuff further in.

Interval 510
First planet! Incredibly small, though, and frozen. It would be lonely there if it didn't have that little moon for company, but it's so far from its star it must take about 250 years to go once around LC 852048.

Interval 511
Now we're moving into the middle of things. Ahead are two frozen gas giants, about 48,000 km across (30,000 miles) we estimate. Difficult to see much on the outer one because it's a long way off our track. Rings? And a couple of moons. The inner planet has been tipped up on its side—its rings circle from top to bottom instead of across the middle. Five moons; and what an eerie green!

Interval 512
We think that this must be the showpiece of the LC 852048 system: in fact, we're making a detour just to look. It's the biggest gas giant yet, surrounded by vast rings measuring 280,000 km (175,000 miles) from tip to tip. Over a dozen moons, and a disc swathed in bands of white and yellow clouds.

Interval 513
What a monster! We're steering past the biggest one yet, but we daren't get too close: our instruments are almost saturated by a huge magnetic field whose trapped radiation could be lethal to *Starbreaker*—and its crew. Incredible, though: a vast, whirling ball of turbulent gases almost 144,000 km (90,000 miles) across, prickling with violent lightning strokes and squally storms. We stopped counting moons when we got to sixteen—and it's got rings as well!

Interval 514
We're cautiously navigating a swirling band of debris between The Monster and the red planet ahead—a pity it's a bit off our track. Some of these rocks are as big as 800 km (500 miles) across, but we're more worried about the smaller ones scoring a direct hit. We're working on the theory that these are bits of a small planet which somehow failed to knit together.

Interval 515
Past that small red planet, and on toward the central star LC 852048—or Sol, as we call him now. Now that we can see him as a disc, he has developed a real character of his own: starspots, prominences and flares. It's all mild stuff, though. We might look at Red later, but we're heading straight at Blue Planet. Without doubt this is where the signals are coming from.

Interval 516
It's no distance at all from Red to Blue. This close to Blue, we're absolutely sure it's our planet—or two. It's more like a double planet, with a moon over a quarter of its size! However, we've just spotted a couple of planets closer in to Sol, so we will check over the whole system first before returning to Blue.

Interval 517
There's no chance of life here beyond Blue. The first planet we met was a swirling gas inferno. It definitely had a solid surface, but we reckon that anyone landing there would be roasted, crushed, corroded and suffocated by the extreme conditions. The one nearest Sol turned out to be small, scorched and heavily cratered.

Interval 518
We are bound for Blue Planet orbit injection.

▼ The outer planets in close up.

▲ Sol's smallest and most distant planet. It has a tiny moon for company.

▲ First of the gas giants: a huge, ghostly blue-green world.

▲ First of the ringed planets; but why does it rotate on its side?

▲ The most beautiful planet of all, encircled by vast rings.

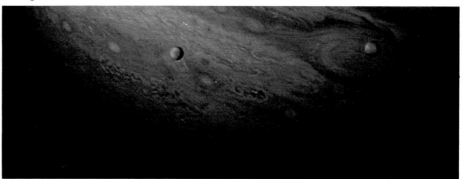

▲ This is the monster, Sol's biggest world. This close-up view shows two of its biggest moons hovering over a churning ocean of poisonous storm clouds.

The Blue Planet

Interval 519
Safe injection: *Starbreaker* is now in orbit around Blue Planet, and we're all set to do a preliminary reconnaissance. Look at that moon: it's huge, and so close! Its gravity must raise tremendous tides on Blue, but it hasn't come off too well itself—that airless, barren surface has been completely battered by meteorites.

What a contrast Blue is! Its atmosphere—just thick enough to shield it from the worst elements of space—is filled with beautiful swirling clouds, and we can glimpse the surface through the gaps. We're beginning to realize why Blue *is* blue. It appears to be two-thirds covered in blue liquid!

Interval 520
We've made ten circuits now and are moving into a lower orbit. That liquid is water. We're beginning to think that if there is—or was—life on Blue, it must be under that water, because what little land there is seems most unstable. You can see from the jigsaw-piece shapes of the land masses that they are still drifting across the globe, and that the crust must crack further in places. Dangerous! Well, "they" must live under that water: but how are we going to contact them?

Interval 521
Into very low orbit now, and there's no sign of life even this close. But we're convinced that Blue is inhabited: our radio receivers are being swamped with signals—whistlings, twitterings, chatterings and some really horrible noises. It must be "their" communications network. Someone joked that it had to be their entertainment channel, but no one in the Universe could be amused by that racket!

Interval 522
We were wrong about the water. "They" *are* land-dwellers. We can now see glowing patches on the night-side of Blue, which must be the lights of cities. We checked on the infrascope, which would reveal collections of buildings as places where the temperature was a bit higher: and sure enough, the cities stood out. There are *thousands* of them.

Interval 523
There's a lot of interference, but we've pinpointed a signal very similar to the one which brought us here. We still can't decipher it. But at least we can tell that it's coming from a small island near the middle of Blue. We're going down to see what sent it. . . .

▲ One of Blue's small islands can be seen from close orbit, but even at this distance there appears to be no sign of life on its surface.

▲ This is the region beaming messages—our landing target on Blue.

▼ City on Blue! Infrascope view reveals buildings (blue) and parks (red).

2: The Solar System

Having completed our fictional journey through the planetary system, let's now take a look at how the Sun and its planets were formed.

By the time our Sun first began to shine about 4.7 thousand million years ago, the Universe was already almost three times as old. Within the great star system to which we belong—the Milky Way—many millions of short-lived massive stars had shone and died. They disappeared in huge explosions which filled the Galaxy with vast clouds of gas and dust. These clouds contained the first heavy elements, which were generated in the nuclear cores of the early stars.

Perhaps it was the explosion of such a massive star which started the slow collapse of the cloud of gas and dust from which our Sun was to form. In addition to abundant hydrogen and helium, the cloud was filled with the debris from our Galaxy's explosive past.

As the clouds collapsed, swirled and heated, the central region got so hot and dense that nuclear fusion began. Thus, 4.7 billion years ago the Sun shone for the very first time.

The exact process of planetary formation is not yet clear, but astronomers believe that the early Sun was still surrounded by a dense cloud of material. Over a very long period small particles of dust, rock and metal collided, crashed and stuck together. Soon larger and larger fragments coalesced and began attracting more and more material to this hardening core.

In this manner objects tens, hundreds and thousands of km in diameter came into being, and gradually their gravitational fields swept up much of the remaining material. The crater-scarred surfaces of Mercury, Venus and Mars and the moons of Jupiter clearly show the dramatic results of this clean-up.

The Sun was also helping in this process. As it grew stronger and hotter, it drove the light gases out of the system. Thus the inner planets contain most of the heavy elements, whereas the outer colder worlds were left with the lighter materials such as hydrogen and helium. Finally, all that remained were the frozen lumps of ice and dust which we today call comets.

The atmosphere forms

Meanwhile in the rocky planetary cores radioactive elements caused immense heat to be released. Soon the light rocks rose upward forming a planetary crust. Volcanoes emerged, and their eruptions, together with outgassing from the interiors, formed the atmospheres of the planets. The smaller bodies like Mercury and the Moon could not hold on to their atmospheres and so gradually became dry and barren.

It was on the third planet from the Sun that life evolved, first in the oceans and then on land. Finally there evolved a being who looked into the sky, saw the stars, Sun and planets and gave them the names we use today.

That being is MAN and this is our Solar System.

◀ Long before the Sun and its family came into being, the Universe was born. Out of the fury of the Big Bang, stars and galaxies slowly formed.

▶ Pounded by thousands of meteorites and racked by volcanic eruptions, this is what Earth would have looked like 4,000 million years ago. Only when the planet had cooled enough could life get under way.

▼ Birth of the Solar System: a dense cloud of dust and gas swirls around the collapsing ball which one day will become our Sun.

Mercury: planet nearest the Sun

The name of the fleet-footed messenger of the gods is very appropriate for this speedy and elusive little world.

Mercury is not much bigger than Earth's Moon, and it never strays far from the glare of the Sun. It is one of the most difficult planets to observe. It is said that many famous astronomers have lived and died without ever having glimpsed Mercury.

To the unaided eye it appears as a bright, slightly pinkish star, sometimes seen in the sunset glow or pre-dawn sky. Through a telescope one can see a small disc, which goes through a series of phases, just like those of the Moon. These changes of appearance are caused by Mercury's revolution about the Sun, once in every 88 days.

Mercury's orbit

Even the world's largest telescopes show little detail on Mercury, and for a long time there was almost no information about its surface. For many years it was even believed that Mercury had a captured rotation, namely that like the Moon does to Earth, Mercury kept one face towards the Sun at all times.

Then with the help of radar observation of the planets, it was discovered that Mercury does in fact rotate on its axis once in every 59 days in respect to the stars.

With respect to the Sun, however, Mercury can be said to have a true day of 176 days. This is because Mercury spins on its axis one and a half times on each orbit of the Sun. This means that to go from noon to noon, Mercury must complete two orbits, spinning three times and taking 176 days to complete this special day.

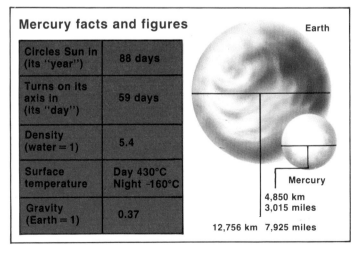

◄ Accelerated by its close approach to Venus, Mariner 10 soon encountered Mercury and provided our first views of this tiny baked planet, closest to the Sun.

Mercury facts and figures

Circles Sun in (its "year")	88 days
Turns on its axis in (its "day")	59 days
Density (water = 1)	5.4
Surface temperature	Day 430°C Night −160°C
Gravity (Earth = 1)	0.37

Earth 12,756 km 7,925 miles
Mercury 4,850 km 3,015 miles

▲ This Mariner 10 shot shows Mercury's crater-scarred and wrinkled surface, where the temperature by day soars to an incredible 430°C.

Even more exotic is the fact that from certain regions on Mercury the Sun would appear to perform a solar loop, the Sun taking eight Earth days first to slow to a stop in the sky, go backwards, and then resume its journey across the sky. When the planet is at its closest to the Sun, its orbital speed is faster than its rotational speed, hence the Sun's *retrograde* loop in the Mercurian sky. This peculiarity is caused by the fact that Mercury's path of orbit around the Sun is not exactly circular; it is in fact one of the most eccentric of all the planets. At its closest Mercury approaches to within 46 million km (29 million miles) of the Sun, and at its most distant strays to 70 million km (44 million miles).

Mercury revealed

Until 1974 the only clues as to the nature of Mercury's surface were a few dusky markings reported by telescopic observers from time to time. Then in 1974 Mariner 10, having flown past Venus, made a series of close approaches to Mercury, revealing a crater-strewn world, dead and wrinkled like a dried fruit.

The photographs returned by Mariner look remarkably like our Moon, with a massive number of craters scarring the surface. Certain areas, less pock-marked than others, have been called intercrater plains. These would appear to be regions where the surface has gone through a period of heating and softening. Later, as the surface cooled, shrunk and hardened, great ridges were formed, sometimes as high as 3 km (2 miles) and hundreds of km long.

Future space tourists who flock to Mercury to see the Solar loop in the sky had better be well prepared for this dried-up, dead little world destined to remain barren and lifeless for ever.

Venus, the hell world

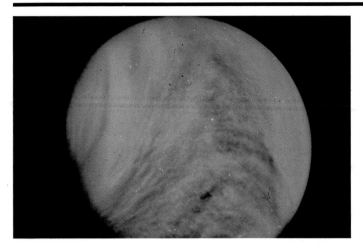

◀ The surface of Venus is hidden from view by a dense mantle of clouds encircling the planet.

Orbiting the Sun at an average distance of 108.2 million km (67 million miles), Venus is the second closest planet to the Sun and circles between the orbits of Mercury and Earth.

Like Mercury, Venus shows phases going from new to full during a cycle, as it completes one orbit every 225 days. This pattern was first observed by Galileo, and he used it to prove that the planets revolved around the Sun, not the Earth.

Unlike Mercury, however, Venus is easy to spot in the sky, since it is often the brightest object to be seen, apart from the Sun and the Moon, and resembles a brilliant lamp in the sky.

There are two reasons for this. First, Venus comes closer to Earth than any other body with the exception of the Moon. At its closest it is a mere 38 million km (24 million miles) away.

Second, Venus is covered by a dense layer of cloud which reflects sunlight back into space. The cloud also hides the surface of the planet from view, which has always been a great disappointment to astronomers, since in size Venus is very similar to our own planet and has often been called the Earth's twin.

Venus is 12,100 km (7,520 miles) in diameter compared to Earth's 12,756 km (7,926 miles), but this is where the similarity ends. Unlike Earth, a description of Venus sounds more like a vision of Hell than anywhere else. Here we have a world shrouded by clouds of carbon dioxide, with a surface hot enough to melt lead.

All this information about Venus has only been discovered in the past twenty years, largely by using radio astronomy, bouncing radar from the surface, and direct exploration by planetary probes. Earth-based optical telescopes can show nothing of the surface, but photographs from spacecraft in ultraviolet light have revealed strange and beautiful weather patterns.

The planet itself spins very slowly, rotating once in 243 days, so that a "day" on Venus is longer than its year. In addition, the spin is retrograde, i.e. from east to west, which is quite different from all the other major planets. Possibly this slow rotation is the reason for the lack of any magnetic field.

Why Venus should be so different from Earth is still something of a mystery, but it seems fairly certain that its high surface temperature is due to the greenhouse effect of the atmosphere. It appears that the dense carbon-dioxide atmosphere allows sunlight to heat the surface, but traps the outgoing heat radiation, resulting in a surface temperature of 470°C.

Venus in close up

In recent years a number of spacecraft have landed on and orbited Venus. From their reports we can imagine what it would be like to drop through the clouds and explore this curious world.

From high above the planet all we see are brilliant yellowish clouds. Having passed through the shock wave in the solar wind 7,400 km (4,600 miles) out, followed by a turbulent region in the ionosphere at 390 km (240 miles), we begin our descent through the thin upper atmosphere.

At about 145 km (90 miles) the clouds appear as a smoglike haze and the braking forces are beginning to feel quite strong. At 65 km (40 miles) the Sun begins to dim and only 3 km (2 miles) deeper the Sun is lost behind a thick, diffuse layer of yellow sulphurous clouds. Peering out we can only see to a distance of 6 km (4 miles) through the tiny particles of sulphuric-acid smog. Our instruments report that the outside temperature is 13°C and the pressure half that which we experience on the Earth's surface.

Venus facts and figures	
Circles Sun in (its "year")	225 days
Turns on its axis in (its "day")	243 days
Density (water = 1)	5.2
Surface temperature	470°C
Gravity (Earth = 1)	0.9

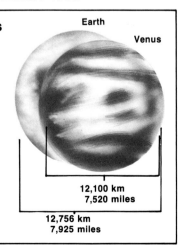

12,100 km / 7,520 miles
12,756 km / 7,925 miles

▼ Radar map of Venus's surface. Mountains show red.

▲ Below the clouds: Venus's highest mountain, Maxwell.

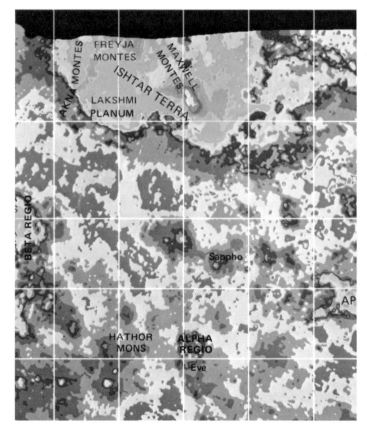

By now we are in a pleasant glide, the dense atmosphere having reduced our speed so well. It gets increasingly dark and our vision is limited to 1½ km (1 mile). Suddenly we break through the upper cloud layer and a beautiful vista opens up clouds above and below. As we pass through the lower clouds, the temperature continues to rise and is now over 200°C. Atmospheric pressure is equal to that on Earth.

The sulphur clouds are extremely dense here, and we cannot see very far. The temperature and pressure continue to rise and then at about 30 km (19 miles) we are below the main clouds. Visibility is about 80 km (50 miles), but all we can see is a dull, reddish glow. By 19 km (12 miles) the outside temperature is 380°C and at 6 km (4 miles) high we can glimpse some craters and mountains shimmering in the dull red murk.

Now we're at ground level the temperature is 430°C and the atmospheric pressure 90 times that on Earth's surface. Everything is a dull red and we cannot see further than 2–3 km (1¼–2 miles). The thick atmosphere completely distorts our view, and because of this we seem to be in a great bowl-like depression, with the horizon curving up to meet the sky.

Someday this will be the view which will confront visitors from Earth. Until then we must be content with the views from our robot explorers.

Planet Earth, Man's home

◀ The Earth, photographed by astronauts aboard Apollo 17. The cloud systems, the oceans and the African continent show up well.
▼▶ From far out in space, the Earth is just another planet. Here we see it going through its phases from crescent to full and back again.

7:30 A.M.

3:30 P.M.

NOON

Man has always found it difficult to think of his world as a mere planet. For thousands of years it was believed that the Earth lay at the heart of the Universe and it was only with the greatest difficulty that the observations which proved otherwise were accepted.

Earth, third planet from the Sun, orbits at an average distance of 149.7 million km (93 million miles) once every 365.25 days—a period of time which we call a year. Spinning on its axis once in 24 hours, Earth rotates west to east. This causes the Sun to rise in the east and set in the west, providing a regular and convenient celestial clock.

Apart from the apparent movement of the Sun and stars, there are few clues to tell us that we live on the surface of a rapidly spinning globe, a globe which is whizzing around the Sun at a speed of 96,000 km/h (60,000 mph) and being trailed by the Sun around our galaxy at a speed of 19 km/s (12 mps).

The view from space

To many the first realization that Earth was a planet came when views of our world were returned by spacecraft. Now at last the 12,756 km (7,927 miles) globe of Earth actually looked like a planet, a beautiful blue planet with pale brown continents hiding under the white wisps of the weather systems.

For the first time the phases of Earth were seen as the ATS satellite, perched 36,000 km (22,500 miles) above the surface, showed the slender crescent Earth of morning, the full Earth of noon, and the waning crescent of nightfall.

Earth is the largest of the inner planets and in comparison with the others has much to distinguish it. There are large tracts of water, the oceans which cover most of the surface, and a dense, oxygen-rich atmosphere. It is also the only planet which we know to support life.

Our protective shell

Originally the Earth had an atmosphere made up of hydrogen and helium. Gradually, outgassing from rocks and volcanoes, and then photosynthesis in plants (a process in which carbon dioxide is taken in and oxygen released), changed its composition to that which we have today. At present 78 per cent of the atmosphere is nitrogen, 20 per cent oxygen and the balance is made up of such gases as argon, carbon dioxide and neon.

The bulk of the atmosphere clings within 8 km (5 miles) of the Earth, but it extends outward, gradually getting thinner, as far as 500 km (300 miles), well into the region traversed by many spacecraft.

Earth has a powerful magnetic field, probably generated by a dynamo effect deep within the planet's molten core. Extending far into space, this field interacts with the solar wind—a stream of charged particles constantly being shot out from the Sun. The combination of magnetosphere and atmosphere provides a very effective shield, preventing harmful radiation penetrating to the planet's surface.

Far above Earth, at an average distance of 384,400 km (239,000 miles) is the Moon, which orbits

10:30 A.M.

7:30 P.M.

▲ The beautiful surface of planet Earth, sculpted by the forces of mountain-building, wind, water—and Man himself.

▲ Earth's companion in space, the Moon. Most of this view shows the far side, hidden from Earth.

astronomers, seeing the dark regions, called them Maria, thinking that they were like seas and oceans on Earth. These areas were also heavily cratered in the past, but dark lava burst through the thin lunar crust thousands of millions of years ago. The crust on the far side of the Moon, not visible from Earth, is also cratered, but the surface seems to have been much stronger, for there are few dark regions.

The most easily observable effects of the Moon on Earth are the tides, the rising and falling of the oceans caused by Moon's gravitational attraction.

As yet the Moon is the only world apart from Earth which man has visited in person. At the moment it waits, lonely and desolate, for man's return to its barren gray surface.

Earth once in every 27.3 days. Actually the Moon is so big, relative to Earth, that in some ways Earth/Moon could be thought of as a double planet.

The gravitational pull of the Earth has slowed the Moon's rotation, so that it now takes 27.3 days to rotate once, the same time it takes to circle the Earth. The result of this is that one face of the Moon always points toward Earth.

To the unaided eye the Moon is covered with light and dark patches. A pair of binoculars clearly shows the heavily cratered, bright highland areas and the dark, so-called seas. Of course, there are no seas on the Moon—it is an airless waterless world—but the early

Earth facts and figures

Circles Sun in (its "year")	365.25 days
Turns on its axis in (its "day")	23 hr 56 min
Density (water = 1)	5.52
Surface temperature	22°C
Gravity	1

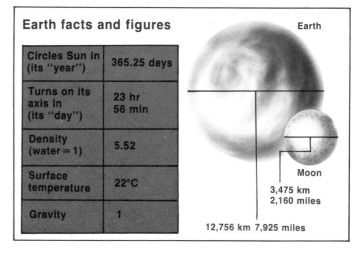

Earth

Moon

3,475 km
2,160 miles

12,756 km 7,925 miles

The red planet

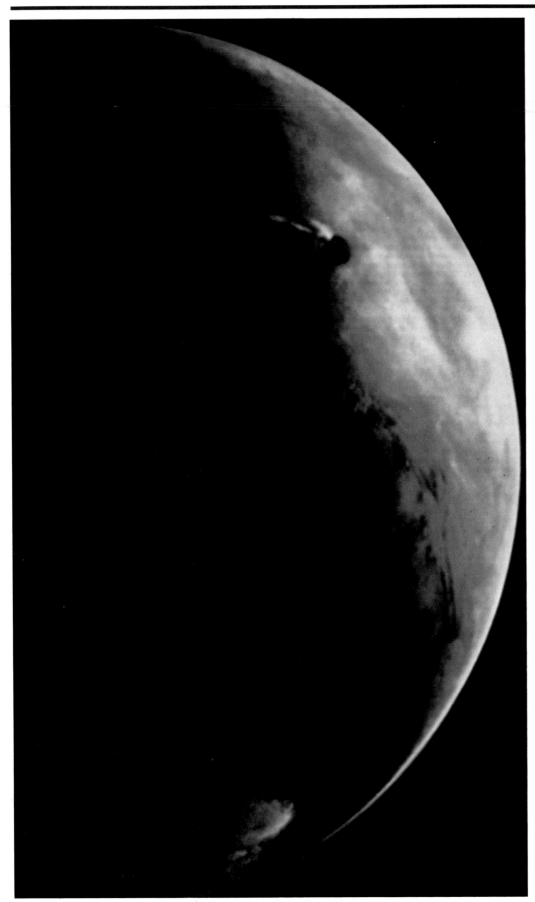

▲ Phobos, 22 km (13½ miles) in diameter, is Mars' largest moon.

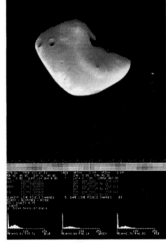

▲ Like its companion, Deimos looks just like a cosmic potato.
◄ Viking's view of the crescent Mars with clouds atop its highest volcano.

Mars facts and figures	
Circles Sun in (its "year")	687 days
Turns on its axis in (its "day")	24 hr 37.5 min
Density (water = 1)	3.9
Surface temperature	−23°C
Gravity (Earth = 1)	0.38

▲ The Red Planet, seen by the Viking lander as it combed the soil for traces of life.

Mars, the fourth planet from the Sun, is one of the most easily identifiable to the unaided eye. At its closest and brightest Mars resembles a brilliant red jewel gleaming among the stars.

Mars orbits the Sun at an average distance of 227.8 million km (141.6 million miles), taking 687 Earth days to complete one circuit. With an equatorial diameter of 6,787 km (4,217 miles), Mars is just over half the size of the Earth.

At first glance, Mars seems to be somewhat similar to our own planet. It rotates on its axis in 24.5 hours, a Martian day being very similar in length to one on Earth. It has a thin atmosphere, and dark markings on its surface. The two white polar caps change shape as they melt in the feeble warmth of Martian summer, only to reform again during the winter.

Life on Mars?

It is no wonder then that Mars has held such a fascination for Earthlings—its pink disc, dark areas and polar caps can be seen even in small telescopes. For many years there was great speculation about the possibility of life existing on Mars, and some believed that Mars was the home of alien beings with envious eyes turned toward our planet Earth.

However, now we realize there is no chance of that, for in recent years Mars has been mapped and visited by a series of orbiter and lander spacecraft.

The Mars revealed by the Viking spacecraft is a truly spectacular world. Its surface is crater-scarred like the other inner planets, but it is covered with tremendous features, giant volcanoes, a huge canyon and what would appear to be dried-up river beds.

The probes also revealed that the Martian atmosphere is extremely thin. At surface level the atmosphere is as thin as the Earth's at a height of 32 km (20 miles). The Martian air is largely composed of carbon dioxide (96 per cent), nitrogen (2.5 per cent) and argon.

Each landing craft had a scoop to collect samples of Martian soil. In a special incubator carried aboard, a series of biological tests were conducted to search for traces of life. Nothing was found, although a series of chemical tests provided somewhat ambiguous results, so perhaps the Martian soil is not as dead as it appears.

The pictures returned showed vistas of scattered rocks, sand dunes and distant hills, sometimes covered in early morning frost, but the weather reports indicated that the temperature rises little during the day. The pinkish tinge of the sky comes from the small dust particles carried aloft. The soil itself is red, which would appear to be due to an abundance of iron oxides (rust) in the surface materials.

The most puzzling aspects of Mars have been the huge areas apparently shaped by flowing water. Even though it is now known that the polar caps are a mixture of water, ice and carbon dioxide, there is not enough there to account for the amount of erosion.

Some scientists believe that frozen beneath the Martian surface is a huge amount of water. It has even been suggested that over a period of thousands of years the inclination of Mars in its orbit changes. The result would be the unlocking of water frozen in the polar caps, and possibly from beneath the planet's surface.

Jupiter, the giant world

At 778 million km (483.5 million miles) from the Sun we encounter Jupiter, the first of the outer planets, which orbits the Sun once every 11.8 years.

Jupiter is much more than a planet: it is a huge globe of gas 11 times the diameter of Earth, weighing two and a half times as much as all the other planets put together. Orbited by a family of 16 moons and radiating over twice as much heat as it gets from the Sun, Jupiter and its family could rightly be thought of as a mini-solar system.

Four of Jupiter's moons, or satellites, are easily visible with a pair of binoculars and their orbital motion is clearly noticeable after a short period of time.

Most amateur telescopes show the 142,800 km (88,750 mile) globe as having a rather squashed appearance and crossed by light and dark belts which move across the planet with various whirls and spots. The flattened appearance of Jupiter is caused by its very high rotation speed of just under 10 hours, the fastest of any planet.

A prominent feature in Jupiter's atmosphere is the great red spot, a huge oval disturbance the width of three Earths. It has been observed from Earth for hundreds of years and has recently been revealed to be a huge hurricane-like storm in Jupiter's atmosphere.

The Voyager mission

In 1979 two American spacecraft, Voyagers I and II, flew past Jupiter and a number of its satellites, leaving in their wake a trail of new discoveries which delighted scientists and fascinated onlookers.

The Voyagers returned stunning photographs of Jupiter and its satellites. Now at last the belts and weather systems could be seen in great detail.

The great bands of cloud were found to be zones of hot and cold gas. These rise and fall as first they are heated in the planet's interior and rise to the surface, appearing as the lighter zones. The darker belts are the cool gases descending once more.

The temperature and chemical range on Jupiter is immense. At the top of the cloud layer it is as cold as −150°C, while lower down in the clouds it gets as hot as +75°C.

Unlike planet Earth, which has only one zone of weather, Jupiter has three—one of water, one of ammonium hydrosulphide and one of ammonia. Below this atmosphere it is mainly liquid hydrogen and helium.

Deep within, Jupiter probably has a molten core of silicates and metals the size of the Earth. This core is surrounded by a shell of liquid hydrogen. Under the great temperature and pressure this form of hydrogen is known as metallic and is an electrical conductor. As this liquid circulates, it acts like a huge dynamo, creating a magnetic field around Jupiter.

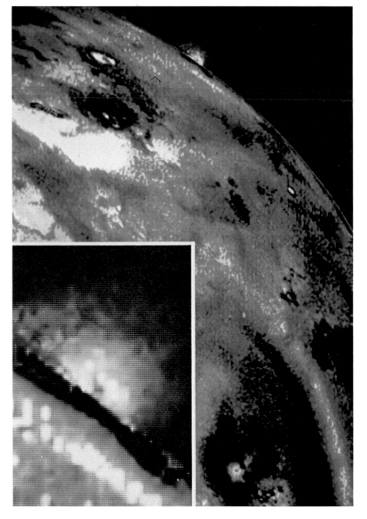

▲ Sulphur-swathed Io, one of Jupiter's largest moons, is peppered with active volcanoes (see inset). The eruptions are continually changing its surface.

Jupiter's magnetosphere extends far beyond the planet and is much bigger than the Sun, entrapping countless thousands of millions of high-energy particles in invisible radiation belts. All of Jupiter's moons are embedded within the magnetosphere and are bathed in radiation so intense as to be lethal to man. Callisto, the most distant large moon, is perhaps the only one our futuristic tourists might attempt to visit.

One of the surprise discoveries of Voyager was a ring of material extending 58,000 km (36,000 miles) from Jupiter's cloud tops. Composed of very small particles, too small to be individually visible, the ring can be seen when illuminated by sunlight from behind.

Next came the photographs of Jupiter's satellites.

▲ A montage (not to scale) showing Jupiter and its four largest moons. Giant gasbag Jupiter is so massive that it holds on to at least 16 moons.

▲ Jupiter's main moons—Europa, Callisto, Ganymede and Io (clockwise from top left). Their varied surfaces will present explorers with a fascinating puzzle.

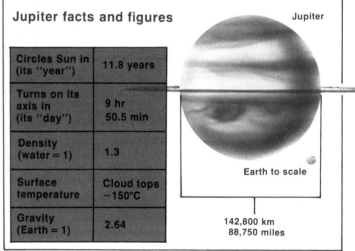

Jupiter facts and figures

Circles Sun in (its "year")	11.8 years
Turns on its axis in (its "day")	9 hr 50.5 min
Density (water = 1)	1.3
Surface temperature	Cloud tops −150°C
Gravity (Earth = 1)	2.64

142,800 km
88,750 miles

The most interesting were those of the four largest—Io, Europa, Ganymede and Callisto—large enough almost to be planets in their own right.

The satellites of Jupiter

Io, the nearest of the satellites, is spectacularly vivid, its 3,640 km (2,260 miles) globe being covered in red, orange and yellow splotches. Orbiting at an average of 421,600 km (262,000 miles) from Jupiter, Io is pulled and tugged by the gravity of Jupiter, Europa and Ganymede. Its crust seethes with tidal forces which heat the upper layer, producing huge volcanic eruptions.

Slightly smaller than Io, Europa, 3,130 km (1,945 miles) in diameter, resembles a huge cracked billiard ball. Europa's rocky interior is covered by a thick layer of rather dirty ice. The streaks which cover the surface seem to be cracks filled with upwelling water or ice. Few impact craters are visible, indicating a young surface.

Orbiting Jupiter at a distance of 1,070,000 km (665,000 miles), Ganymede is 5,270 km (3,275 miles) in diameter. It is covered by a very thick layer of ice which on the surface has been pulverized, crushed and darkened by millions of micrometeorites. Here and there brilliant white splashes mark the impact of some asteroid millions of years ago. Unlike Europa, Ganymede's surface is not smooth. There are many rounded crater rims and chains of ice ridges about 100 km (60 miles) high and up to 10 km (6 miles) long.

Callisto, 4,850 km (3,010 miles) in diameter, circles Jupiter at a distance of 1,880,000 km (1,168,000 miles). Callisto has a structure similar to Ganymede, but its ice surface bears the scar of a huge impact thousands of millions of years ago. This blasted a hole 200 km (125 miles) across and many km deep.

Saturn the beautiful

Undoubtedly the highlight of future Solar System tours will be Saturn, perhaps the most beautiful of all the planets.

Saturn is the second largest planet in the Solar System. It is 120,000 km (75,000 miles) in diameter, rotates once in 10.5 hours and orbits the Sun at an average distance of 1,427,000 km (887,000 miles) taking 29.5 years to complete one Solar journey.

Like Jupiter, Saturn is a largely gaseous planet, composed of a foggy mixture of hydrogen, methane and ammonia. Once again it is the high rotation speed that causes the belts and zones which can be seen crossing its surface.

The atmosphere of Saturn is not as vivid or as active as that of Jupiter, but it does share certain features such as spots, bands and other markings.

A heat-detecting instrument carried by Pioneer 11 during its Saturn flyby of 1979 found a good variation of temperature across the surface of the planet. The average cloud-top temperature is around −130°C and in general the darker belts are slightly cooler than the warmer bright zones.

The rings of Saturn

Of course, Saturn's most spectacular asset is its magnificent ring system. Early astronomers with small telescopes could not see the rings as such and were greatly puzzled by Saturn's changing appearance. Over a period of years we see the rings from different angles, and often when seen edge-on they seem to disappear altogether.

This shows how thin the rings are, perhaps just a few km at the most. What they lack in depth the rings make up for in width, encircling their planet to a width of 275,000 km (171,000 miles). Recent Pioneer 11 observation suggests that the rings are composed of many very tiny fragments of ice-covered dust, which circle Saturn like billions of tiny moons. From close up they must present a spectacular sight as they glitter like countless diamonds sparkling in the Sun.

Future explorers had better be careful, for Saturn, like Jupiter, has a magnetic field and radiation belts. Although much weaker than Jupiter's, and not strong enough to affect automatic probes, radiation levels close to Saturn are still strong enough to kill an unprotected human.

Pioneer 11 was the first spacecraft from Earth to approach Saturn. It did so in 1979 as a bonus from its Jupiter mission of 1974, having journeyed for 6½ years through space checking out the flight path for the more sophisticated Voyager craft which followed it to Jupiter in 1979 and to Saturn in 1980 and 1981.

Saturn's moons

Saturn is believed to have as many as 20 moons, including a 320 km (200 mile) asteroid-sized moon detected by Pioneer 11. One of Saturn's moons, Titan, 5,135 km (3,190 miles) in diameter, is the only moon in the Solar System known to possess a dense atmosphere of nitrogen and other gases. It is known that complex hydrocarbon compounds exist in Titan's atmosphere, while its surface is probably covered by an ocean of liquid nitrogen.

Scientists now believe that the atmosphere on Titan is up to three times as dense as that of Earth. Titan can be thought of as an Earth-like planet, but in deep freeze.

Voyager observations of the other moons reveal an icy composition with complex crater-scarred and dusty surfaces.

The origin of the moons is still unclear, but it is possible that a number of the smaller ones could be the nuclei of comets captured by Saturn in the distant past.

Saturn facts and figures	
Circles Sun in (its "year")	29.5 years
Turns on its axis in (its "day")	10 hr 40 min
Density (water = 1)	0.7
Surface temperature	cloud tops −180°C
Gravity (Earth = 1)	1.15

120,000 km / 75,000 miles — Earth to scale

◄ Voyager 1 surveys Saturn at close quarters during a flyby in November 1980. The probe took 3 years to get there at 40,000 km/h (25,000 mph).

▲ Saturn's beautiful ring system, 275,000 km (171,000 miles) across, is made up of billions of fragments of rock, ice and dust.

Uranus, Neptune and Pluto

Beyond Saturn the Solar System is largely empty space, presided over by three known planets, their satellites, some debris from the planets' formation, and the frozen nuclei of comets.

Uranus

None of the outer planets can be seen with the unaided eye, but the closest to Saturn, Uranus, is visible in small telescopes as a faint greenish star.

Uranus appears green because its atmosphere is composed of a mixture of hydrogen, helium and methane, which absorbs red light.

Beneath the 10,000 km (6,250 mile) deep atmosphere, there is probably a central core about 25,000 km (15,500 miles) across consisting of a layer of ice and molecular hydrogen and surrounding a rocky core about 8,000 km (5,000 miles) in diameter.

Uranus is 50,800 km (31,600 miles) in diameter and orbits the Sun every 84 years at a distance of 2,870 million km (1,783 million miles). Like a planet lying on its side, Uranus is unique in that its axis of rotation lies almost on the plane of its orbit. This results in each pole alternately facing the Sun every 42 years.

Never visible from Earth, and only discovered in 1977 when Uranus passed in front of a star, are the rings of Uranus. Numbering a possible nine, the rings appear to be composed of tiny dark fragments with a similarity to graphite.

Neptune

Neptune, the next planet, orbits the Sun at a distance of 4,500 million km (2,800 million miles) every 165 years. Discovered in 1846 because of its gravitational disturbances of Uranus's orbit, Neptune is a gas giant very similar to Uranus.

Neptune has two satellites, one of which, Triton, perhaps 6,000 km (3,700 miles) in diameter, is the largest satellite in the Solar System.

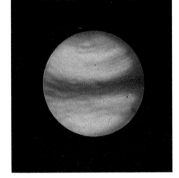

▲ Voyager's 1989 target: unknown Neptune, with its dense bluish clouds.

▲ Pluto's moon revealed! Blurred image shows Charon (top left) and Pluto (right).

Pluto

Normally Pluto is the most distant planet in the Solar System, orbiting the Sun at an average distance of 5,900 million km (3,600 million miles) once every 248.5 years. However, Pluto's orbit is quite elliptical, and at certain times this brings the tiny planet closer to the Sun than Neptune. This will be the case from January 1979 until March 1999.

Pluto was discovered in 1930. Very little is known about it, for it appears as a mere speck of light in even the world's largest telescopes. A great breakthrough came in 1978 with the discovery of a large satellite circling Pluto. Observations of the satellite, subsequently called Charon, have at last settled some of the mysteries about Pluto.

For example, we now know that Pluto is a small planet, possibly a ball of rock and ice covered in methane. It is approximately 3,000 km (1,900 miles) across and rotates once in 6.38 Earth days.

Its satellite, Charon, is 1,200 km (750 miles) across and is only 17,000 km (10,600 miles) away from Pluto. It takes 6.38 days to orbit Pluto and taken together with Pluto both have a mass of 1/600th that of the Earth.

The Pluto/Charon system is the nearest example of a double planet to be found in the Solar System.

◀ Uranus, Voyager's target in 1986. This gas giant is circled by a set of very thin rings, which were discovered because they blotted out the light from distant stars lying behind. They seem to be made up of much tinier particles than those which make up the rings of Saturn. Mysteriously, Uranus circles the Sun tipped up on its side.

Uranus, Neptune and Pluto facts and figures

	Uranus	Neptune	Pluto
Circles Sun in (its "year")	84 years	164.8 years	247.7 years
Turns on its axis in (its "day")	11 hr (uncertain)	16 hr (uncertain)	6 days 9 hr
Density (water = 1)	1.7	1.7	?
Surface temperature	cloud tops −210°C	cloud tops −220°C	cloud tops −230°C
Gravity (Earth = 1)	1.17	1.18	?

Uranus 50,800 km / 31,600 miles
Neptune 48,600 km / 30,200 miles

Comets, meteors and meteorites

A bright comet can be a beautiful and awe-inspiring sight, adorning the skies for days or even weeks. Visually a comet is composed of a bright nucleus from which a huge tail or tails streams millions of km into space.

However, much of this performance is pure show, for even the largest comets could be called bags full of nothing. The central nucleus of even the most brilliant comet is probably just a few km across, a mixture of ice and dust, a giant dirty snowball.

Comets seem to originate from a great swarm of such bodies which orbit the Sun at distances almost halfway from Pluto to the nearest star. On entering the Solar System and approaching the Sun, the ices change into gas, which forms the beautiful tail. Dust particles from the nucleus also form a tail, and sometimes when the two tails are not exactly in line, a beautiful effect is produced.

At its closest to the Sun the comet has a brief moment of glory before rapidly fading as it drifts back into the dark abyss.

Sometimes a comet may pass close to a planet and have its orbit altered to bring it much closer to the Sun. Such regular or periodic comets are rapidly depleted by their Solar approaches, and are not as spectacular or as beautiful as their unexpected relatives.

The one exception is Halley's Comet, which orbits the Sun once every 76 years. Due back again in 1986, it is expected to put on a good show, particularly for observers from the southern hemisphere of Earth.

Meteors

On a good dark night a few minutes' of sky-watching will be rewarded with the sight of a meteor or shooting star. Such objects have nothing to do with the stars themselves, which are huge suns far away in space.

A meteor is caused by a tiny particle, no bigger than the head of a pin, burning up as it plunges Earthward in our dense atmosphere.

Most meteors are sporadic, that is they appear from anywhere in the sky. However, at certain times the Earth, as it circles the Sun, crosses the path once taken by a comet. Comets are very untidy and leave behind them a trail of dust. These dust particles collide with Earth's atmosphere and produce a meteor shower. There are many meteor showers each year and on a good Moon-less night they can provide a spectacular show.

Meteorites

Sometimes larger particles hit the atmosphere and

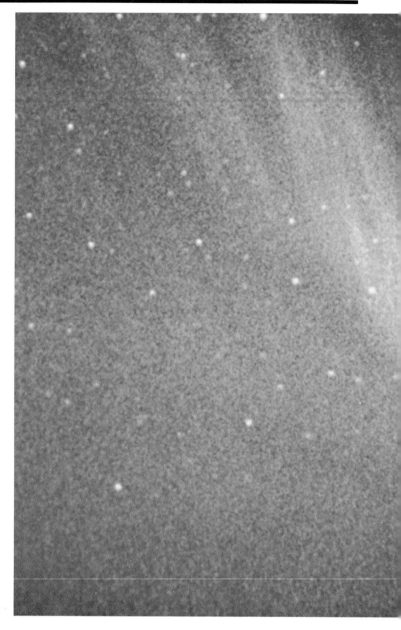

produce super meteors called fireballs. Fireballs can be very spectacular, often leaving behind a trail which can linger high in the sky for many minutes.

When a lump of rock or debris is big enough, it will not completely burn up but will survive its fiery plunge Earthward and land on the Earth's surface. Such objects are known as meteorites. Most meteorites are small, only a few kg in weight.

On very rare occasions a huge meteorite will crash through the Earth's atmosphere and the result is a vast explosion on the Earth's surface, often resulting in the formation of a giant crater. Unlike the Moon or Mercury, which have no atmosphere, many of the Earth's craters have been eroded away, but some still exist.

◄ Comet West hangs in the morning sky, its tail streaming out into space.

▲ Pulled to Earth, a meteoroid meets a fiery death in our atmosphere.

▼ Meteor Crater, Arizona, blasted out by a quarter-million ton meteorite.

▲ A fireball—a super-bright meteor which may survive its fiery journey Earthward.

The most famous is Meteor Crater near Winslow, Arizona. It is 1 km (⅝ mile) in diameter and was blasted out by a huge meteorite about 50,000 years ago.

Asteroids

The asteroids are a band of perhaps 500,000 giant rocks and chunks of material which orbit between Mars and Jupiter. They range in size from Ceres, 1,000 km (625 miles) in diameter, to flying mountains a few km across. Most asteroids are highly irregular in shape and are thought to be composed of very primitive planetary material. A large percentage are very dark and are of a possible carbonaceous composition. Who knows, perhaps someday mining the asteroids will be a space industry and not just a science-fiction writer's dream.

3: The view from Earth

In this chapter we look at some of the more obvious ways in which space affects a small spacecraft called Earth.

It's only in the last twenty years that people have broken free of the Earth and have soared into space. Some of you may be lucky enough to go further and actually visit some of the worlds described in the last chapter. But for many people, space is simply something "up there"—something which somehow influences us in some way.

Like all bodies in space, our spacecraft is spinning. At the equator, Earth is whirling around at over 1,600 km/h (1,000 mph), but its spin is so smooth and the world is so big that we can not feel it. It takes 24 hours to go around once, which we call one day. The day is steadily getting longer—or, to put it another way, the Moon's pull on the Earth is slowing it down—but only by one-thousandth of a second every century! Modern clocks, though, can measure this spin-down.

As Earth spins, the part of the world you live on first faces the Sun and then turns away, causing the alternation of day and night. During the day the Sun seems to rise in the east, and then it is carried across the sky before setting in the west. Ancient astronomers invented the sundial to measure the passing of time during the day.

When we on Earth turn away from the Sun, darkness falls and—if it's a clear night—the stars appear. They are around us all the time, of course, but the daytime glare of the Sun drowns them. As you watch, stars rise slowly in the east, while others set in the west. Astronomers of old mapped the stars and used them to tell time as they moved across the sky. But if you look closely, you will find there's one star which doesn't move. This is the Pole Star. Earth's North Pole, quite coincidentally, points almost directly at this star, and so we turn underneath it. If you can find the Pole Star, you

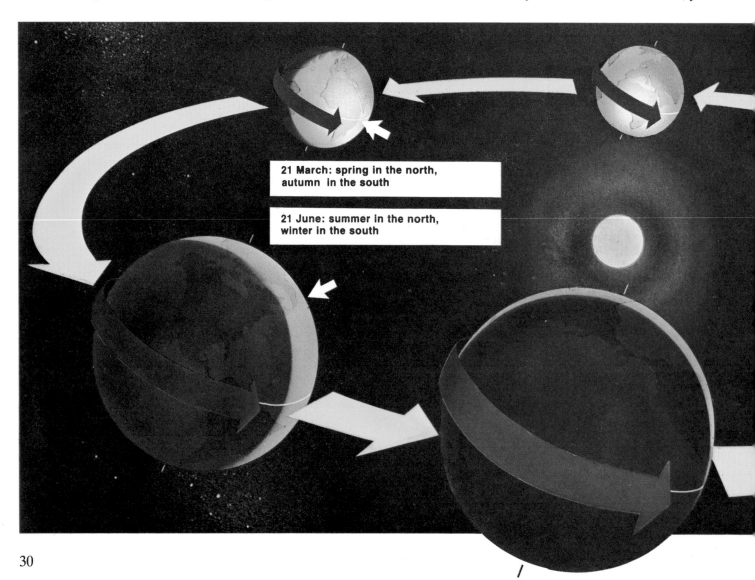

21 March: spring in the north, autumn in the south

21 June: summer in the north, winter in the south

▲ Our spinning Earth, revealed by pointing a camera at the North Pole of the sky and keeping the shutter open for hours. Star trails show our motion.

have automatically found North, so there's no excuse for getting lost in the dark—unless it's cloudy!

Like the other planets, the Earth orbits the Sun. At a speed of 29 km/s (18 mps), the journey takes 365¼ days, or one year, and during this time, we experience changes in weather and in daylight—the seasons.

The seasons

Because Earth spins at an angle, rather than directly upright, northern and southern hemispheres get different shares of the sunlight, depending upon whether they are pointing toward the Sun, away from it, or neither. In December the northern hemisphere is tilted away from the Sun, making the days there short and cold. It is winter in the north: but the southern hemisphere, tilted toward the Sun, enjoys summer. The opposite happens in June, when the northerners have their summer. In March and September, when neither hemisphere is tilted toward the Sun, we have the less extreme seasons of spring and autumn.

As we circle the Sun, we are continually looking out in different directions into space, and so we see constantly changing star-patterns during the course of a year.

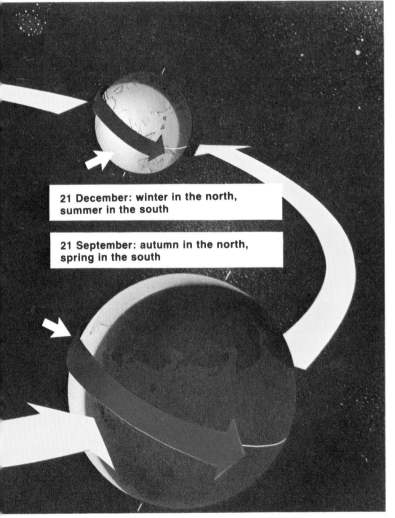

◄ This picture shows the tilted Earth whirling in its orbit around the Sun. As it goes round, each hemisphere tilts more or less toward the Sun, resulting in seasonal changes of temperature.

21 December: winter in the north, summer in the south

21 September: autumn in the north, spring in the south

▲ The gap between the shadows cast by the dolphins' tails tells the time on the dial of this lovely sun-clock at London's National Maritime Museum.

Star spotting: spring and summer

The stars of spring

LOOKING TO THE NORTH
The Plough, or Big Dipper, is overhead and the winter constellations are setting in the west.

The maps on this page show the stars at 22.00 hours local time on 1 March and 1 July.

LOOKING TO THE SOUTH
It's the best time of the year to spot Leo and Virgo. Arcturus (in Boötes, just above Virgo) is very orange, and is the third brightest star in the sky.

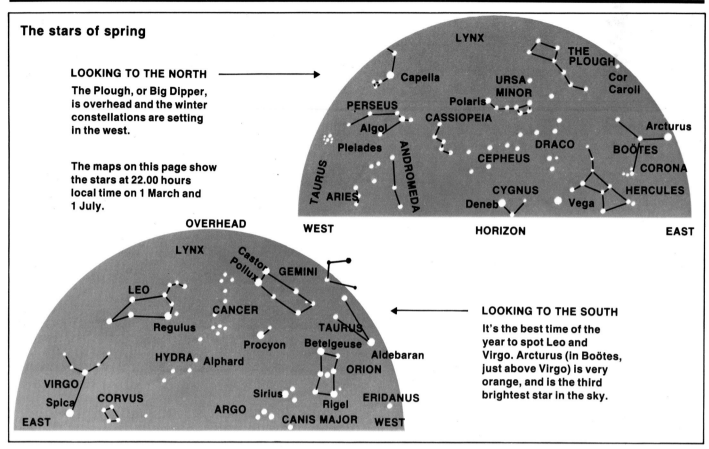

The stars of summer

LOOKING TO THE NORTH
Leo is setting in the west, while the winged horse Pegasus—which simply looks like a large square of stars—makes its appearance in the east.

LOOKING TO THE SOUTH
Blood-red Antares, in the constellation of Scorpius, lies due south. Faint Hercules is high in the sky. Try looking for the globular cluster M 13 described on the page opposite.

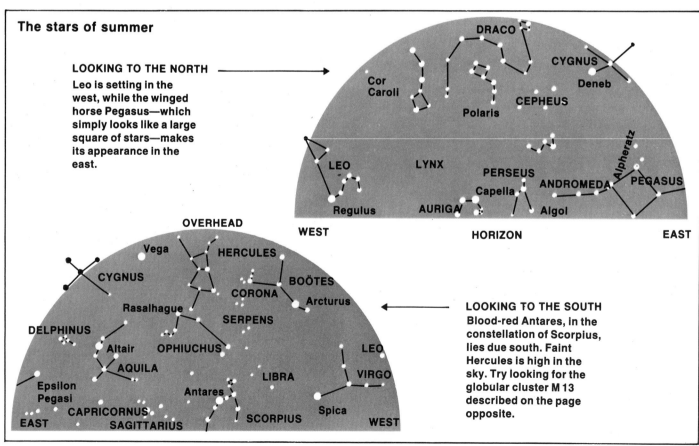

The stars of spring

Thousands of years ago the first astronomers connected up the stars into an array of constellation patterns to help them find their way around the sky. Although it takes a lot of imagination to see the shapes, they named the patterns after animals, birds, people, objects and mythological creatures, and they wove intricate legends around them.

Today we still use the same patterns. Astronomers usually call them by their Latin names, which aren't too difficult once you've learned them. Here are some of the stars and constellations you can look out for during the spring.

The Plough is almost overhead. Find the Pole Star by following the line of the Pointers. Now look at the second star in the Plough's handle (opposite end from Pointers). Mizar is double: it's circled by Alcor, a fainter companion star.

The Plough (sometimes called the Big Dipper or the Saucepan) is actually the end part of a bigger constellation which ancient astronomers called Ursa Major (Great Bear). See if you can pick out the fainter stars which make up the head, nose and paws: the Plough and handle make up the Bear's body and tail. The Pole Star marks the end of the tail of Ursa Minor (Little Bear), which looks like a smaller version of the Plough. Of course, bears don't have tails, and so the old astronomers had to invent some fairly far-fetched legends to account for the tails on the bears in the sky!

Follow the Plough's handle downward to brilliant orange Arcturus in Boötes (Herdsman). Keeps the curve going down to blueish Spica, brightest star of Y-shaped Virgo (Virgin).

Above Virgo, very prominent in the southern sky, is Leo (The Lion). This actually does look like a lion, with a prominent sickle of stars forming the head.

The stars of summer

Vega, Deneb and Altair make a brilliant triangle in the south (The Summer Triangle). Look between these stars for the faint band of the Milky Way, which now stretches right across the sky from north to south.

At this time of year there are actually some constellations which resemble their namesakes! Examples are Cygnus (the Swan: the longest arm of the cross is the outstretched neck); nearby Delphinus (the Dolphin); Corona (the Crown), near Boötes; and Scorpius (the Scorpion), due south now. Blood-red Antares, brightest star in Scorpius, is some 300 times bigger than our Sun.

If it's a really dark, clear night, see if you can spot the globular cluster (M 13) in Hercules. It's made up of nearly a million stars, but looks like a soft, fuzzy patch to the unaided eye. Join up Vega to the middle star in Corona, and you should just see it halfway between.

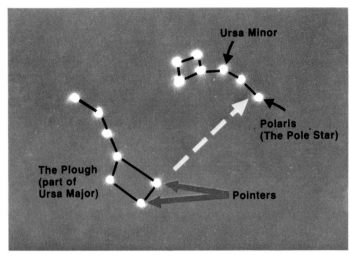

▲ Find the Pole Star by extending the line that joins the Plough's Pointers. The Pole Star always marks north.

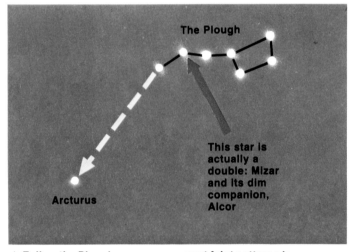

▲ Follow the Plough backward, and you'll find Arcturus. You can easily spot faint patterns in this way if you learn the bright ones first.

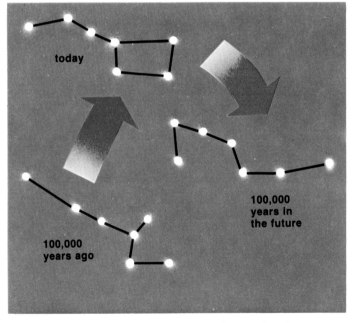

▲ The constellations don't stay the same forever. Over thousands of years, the stars' real motions show up, as we see here with the Plough.

Star spotting: autumn and winter

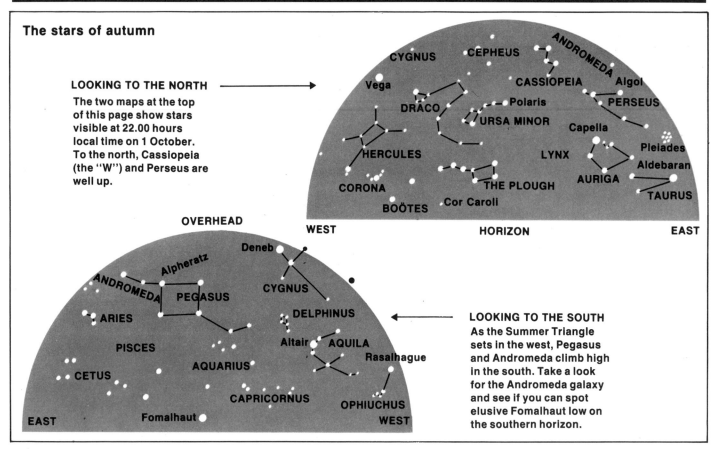

The stars of autumn

LOOKING TO THE NORTH
The two maps at the top of this page show stars visible at 22.00 hours local time on 1 October. To the north, Cassiopeia (the "W") and Perseus are well up.

LOOKING TO THE SOUTH
As the Summer Triangle sets in the west, Pegasus and Andromeda climb high in the south. Take a look for the Andromeda galaxy and see if you can spot elusive Fomalhaut low on the southern horizon.

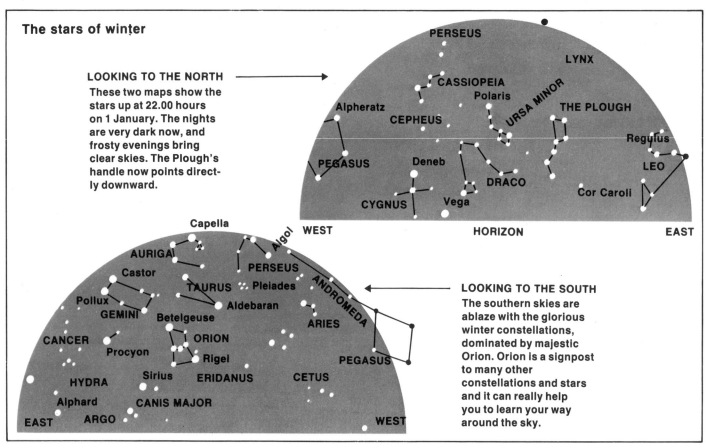

The stars of winter

LOOKING TO THE NORTH
These two maps show the stars up at 22.00 hours on 1 January. The nights are very dark now, and frosty evenings bring clear skies. The Plough's handle now points directly downward.

LOOKING TO THE SOUTH
The southern skies are ablaze with the glorious winter constellations, dominated by majestic Orion. Orion is a signpost to many other constellations and stars and it can really help you to learn your way around the sky.

The stars of autumn

The Summer Triangle is moving into the west, and the sky is rather short on bright constellations. The Plough reaches its lowest point down in the north, but because it lies so close to the Pole, it never sets as seen from the UK and most of the United States. Astronomers say it is *circumpolar*. Stars further across the sky from the Pole Star, like Vega, do pass below the horizon.

On the opposite side of the Pole Star, Cassiopeia (Queen Cassiopeia) is reaching her highest. This W-shaped pattern is supposed to look like a lady in a chair. East of Cassiopeia is the long constellation of Perseus, through which a rich clump of the Milky Way passes. Perseus also contains the variable star Algol, whose name means The Winking Demon. Every 2½ days its brightness falls by half; but Algol's light output is not really changing. It actually consists of two stars—one bright, one dim—orbiting very close to each other. When the dim one passes in front of its bright companion, the light of the whole system is reduced.

A large portion of the southern sky is taken up by the barren square of Pegasus (supposed to look like a winged horse!). East of Pegasus stretches a line of three stars making up Andromeda (the Chained Maiden). Just above the easternmost star lies a misty oval about the size of the full Moon: this is the Andromeda galaxy. At 2¼ million lightyears it's the most distant object visible to the unaided eye.

The stars of winter

This is the best time of year to get to know the sky. The nights are long and dark, and the winter stars are the most dazzling of the year.

Orion (the Hunter) rules the southern sky. It's the most spectacular constellation of all, containing scores of interesting objects. Betelgeuse (The armpit of the Sacred One!) is a slowly-pulsing red giant star over 400 times bigger than the Sun. Hanging below the three belt stars, Orion's sword contains the beautiful, fan-shaped Orion Nebula, a huge gas cloud in which stars have just been born. Follow the belt downwards, and you'll come to Sirius (the Dog Star) of Canis Major (Great Dog), the most brilliant star in the sky.

Following Orion's belt in an upward direction will bring you to orange-red Aldebaran, the eye of Taurus (the Bull). Aldebaran is surrounded by a V-shaped little cluster of stars called the Hyades, which make up the Bull's head. Continue the belt upward beyond Aldebaran, and you'll reach another real cluster of stars, the Pleiades, or Seven Sisters.

East of Orion you will spot yellowish Procyon of Canis Minor (Little Dog), and Castor and Pollux of Gemini (the Heavenly Twins). Brilliant yellow Capella of Auriga (the Charioteer) occupies the overhead spot, where electric-blue Vega shines in the summer.

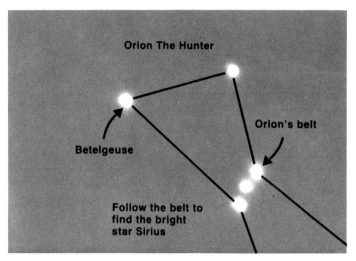

▲ Part of Orion, the most spectacular constellation in the sky.

Betelgeuse is a slowly pulsing red giant star, 400 times bigger than the Sun.

▲ The Pleiades, or Seven Sisters, are a real group of stars. Most people can spot six to ten Pleiads, but there are over 250 stars in the cluster.

Eclipses of the Sun and Moon

To people who lived long ago the Sun and Moon were the most important objects in the sky. The Sun gave light by day; the Moon by night. But every so often something terrible happened: the light went out! Would it be dark forever? How would people protect themselves against the wild animals prowling in the dark? You can imagine how terrifying it must have been for our ancestors when the Sun and Moon were eclipsed.

The early astronomers were hard-headed enough to realize that eclipses were connected with the paths of the Sun and Moon across the sky. They knew that if they measured the changing positions accurately enough, they could forecast when eclipses would happen.

But generally people could depend on the Sun and the Moon, although the Moon's shape appeared to change constantly. Today we know that the cycle of the Moon's phases, from crescent to full and back again, is caused as the Moon reflects different fractions of sunlight back to us as it travels on its monthly orbit about the Earth.

Eclipses happen rarely, for the Moon's orbit is angled to prevent the necessary alignments happening every month. A handful of lunar and solar eclipses occurs each year, and each lasts for a very short time—hours for a lunar eclipse, minutes for an eclipse of the Sun.

And because the alignment needs to be so exact, so-called *total* eclipses are seen over very restricted regions. People just outside the zone will see a less exciting *partial* eclipse.

Eclipses of the Moon are so unsensational that they often go unnoticed. The full Moon simply moves into Earth's shadow and its sunlight is cut off; but its disc

▲ This photograph of Earth, taken during a solar eclipse, shows the Moon's shadow on Earth's surface. Inside the shadow the Sun is blotted out.

never goes completely dark. Usually it glows a dull, coppery red, reflecting sunlight which has been bent around the edge of the Earth by its atmosphere.

Solar eclipses, on the other hand, are so spectacular that astronomers will travel halfway round the world to see them. They are caused when the dark new Moon passes directly in front of the disc of the Sun. By a lucky coincidence the Sun and Moon appear the same size in our sky, and so the overlap is almost exact. Actually the Sun is 400 times bigger than the Moon—but then, it's 400 times further away.

When its brilliant surface is blotted out, the Sun's outer atmosphere—the *corona*—leaps into view as a streaming, pearly halo. The sky all around is dark. Birds and animals sleep. Clouds appear as the temperature drops. It is eerily quiet. And just as suddenly, it is over: the Moon uncovers the Sun, and ghostly night turns into brilliant day.

Modern astronomers are as fascinated by solar eclipses as their long-dead counterparts. Today their task is not to allay the fears of frightened and superstitious people, but to snatch vital information about normally invisible regions of our local star. In those rapidly ticking minutes, you dare not risk an equipment failure!

◀ A total eclipse reveals the Sun's atmosphere: the outer pearly corona, and the inner pink chromosphere.

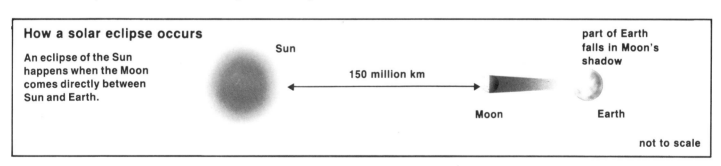

How a solar eclipse occurs

An eclipse of the Sun happens when the Moon comes directly between Sun and Earth.

Sun — 150 million km — Moon — part of Earth falls in Moon's shadow — Earth

not to scale

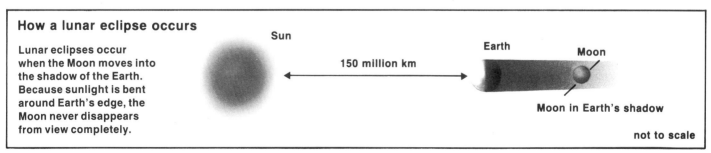

How a lunar eclipse occurs

Lunar eclipses occur when the Moon moves into the shadow of the Earth. Because sunlight is bent around Earth's edge, the Moon never disappears from view completely.

Sun — 150 million km — Earth — Moon

Moon in Earth's shadow

not to scale

Spotting the planets

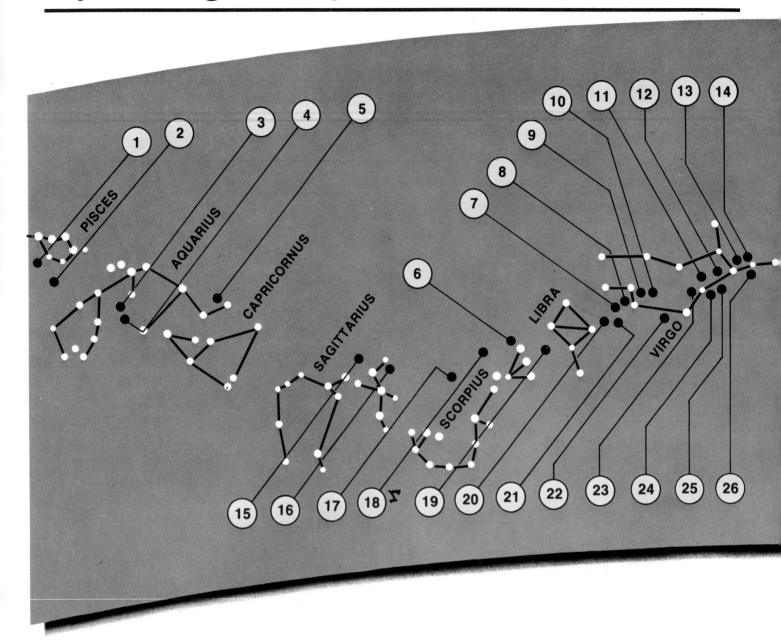

Astronomers have learned to be on the alert whenever a bright planet becomes visible in the sky. It's quite usual for scores of people to phone in reporting their discovery of a supernova (exploding star) or even a flying saucer! It's a mistake which is easy to make, particularly if the sky has been cloudy for weeks and the planet has recently moved into view: but there are a number of ways you can prove that the mystery object is a planet.

First, stars twinkle and planets don't; they shine with a steady glow. Twinkling is simply an effect of living under our sea of atmosphere whose continual churning makes the point-like star images wobble. Planets—close enough to show tiny discs—are unaffected by our air currents. A planet will also give itself away by moving slowly across the night sky over a matter of weeks.

As you would expect, each planet in the night sky has its own character. Mercury, closest planet to the Sun, never strays far from it and can only be glimpsed just after sunset or before sunrise. It looks like a fairly bright star, but because it's always low, it can easily get lost in the haze.

There's no mistaking Venus. Like all planets, it shines by reflected sunlight, but its thick cloud cover makes it an extra-efficient reflector. It's the brightest object in the sky after the Sun and Moon, and it can even cast a shadow on occasions. Look for it after sunset or before sunrise, for, like Mercury, it stays fairly close to the Sun.

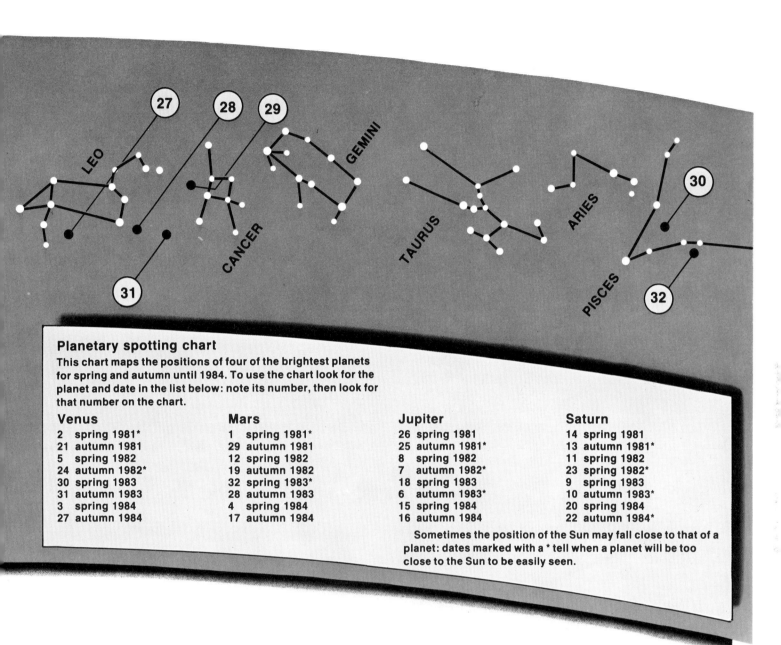

Planetary spotting chart
This chart maps the positions of four of the brightest planets for spring and autumn until 1984. To use the chart look for the planet and date in the list below: note its number, then look for that number on the chart.

Venus	Mars	Jupiter	Saturn
2 spring 1981*	1 spring 1981*	26 spring 1981	14 spring 1981
21 autumn 1981	29 autumn 1981	25 autumn 1981*	13 autumn 1981*
5 spring 1982	12 spring 1982	8 spring 1982	11 spring 1982
24 autumn 1982*	19 autumn 1982	7 autumn 1982*	23 spring 1982*
30 spring 1983	32 spring 1983*	18 spring 1983	9 spring 1983
31 autumn 1983	28 autumn 1983	6 autumn 1983*	10 autumn 1983*
3 spring 1984	4 spring 1984	15 spring 1984	20 spring 1984
27 autumn 1984	17 autumn 1984	16 autumn 1984	22 autumn 1984*

Sometimes the position of the Sun may fall close to that of a planet: dates marked with a * tell when a planet will be too close to the Sun to be easily seen.

Mars shines a brilliant blood-red, and like all the planets outside the orbit of Earth, it isn't restricted to staying close to the Sun in our skies. But in common with all the planets, it does keep to a narrow track in the sky, and this will help you to identify it.

Jupiter comes close to Venus in brightness, shining creamy yellow. A pair of binoculars will just reveal the flattened disc and its four brightest moons.

Saturn moves extremely slowly and is often mistaken for a medium-bright yellowish star.

Uranus can be glimpsed with binoculars if you know exactly where to look. Neptune and Pluto, unfortunately, need the aid of a telescope.

Planets generally move eastward across the sky, but sometimes they turn around and go backwards (*retrograde motion*). Ancient astronomers were puzzled by this, but now we know it happens when Earth overtakes an outer planet, or when an inner planet reaches its extreme position east of the Sun and starts to move back again. It can be entertaining to watch these loops speeded up in a planetarium!

The Sun, Moon and planets all move around the sky in a narrow band called the *ecliptic*, which tells us that the Solar System is a flat disc. The ecliptic passes through the twelve ancient constellations of the *zodiac*, and people used to believe that the positions of the Sun and planets against this starry background could affect our destiny. Today we know that our planets and the distant stars are completely unrelated, and that this practice of astrology can have little or no foundation.

4: Stars

Our local star is the Sun. Although it seems so big and bright to us, it is in fact only an ordinary star.

Pass sunlight through a cheap glass prism and it spreads out into a rainbow. Incredibly, this simple pattern is the key which unlocks the secrets of the Sun and other stars; it tells astronomers what they're made of, how hot they are, how fast they're moving and even how big they are.

First, an astronomer must take the *spectrum* of the Sun or a star. He does this with a sensitive *spectroscope* attached to his telescope, and the resulting photograph looks like a rainbow crossed with dark lines. The rainbow comes from hot gases in the deeper layers of the star; the dark lines from cooler gases near the surface. It is the dark lines which give most of the clues. Each chemical element (e.g. iron, sulphur, oxygen) has its own recognizable pattern of spectral lines, and so by looking for these patterns in a spectrum, an astronomer can disentangle the different elements which are present in that star. The width and darkness of the lines give him more detailed information about the conditions there.

Although our Sun appears so bright and fierce to us, its spectrum reveals that it is only an ordinary star. It is made up of hot gas throughout: three-quarters hydrogen gas, one-quarter helium, and all the other elements are there in tiny amounts. Everything about our Sun is average: its (mighty to us!) size of 1.4 million km (864,000 mile) diameter: its surface temperature (5,800°C); and its brightness.

The Sun has been shining steadily for the past 5,000 million years and seems likely to continue for a similar period in the future. It works like a colossal hydrogen bomb. Deep in the Sun's dense core the temperature is a searing 14 million degrees. Under these extreme conditions a process called *nuclear fusion* welds the Sun's hydrogen into the next heaviest element, helium. Every time this happens, a little bit of mass is converted into energy, which streams out from the heart to keep the Sun shining. Because there is so much matter in the Sun, the amount of matter converted to energy by fusion is a staggering 4 million tons a second!

Our Sun may shine steadily, but this doesn't mean that it's unchanging. The gases which make up its surface are in constant turmoil, for this is where the Sun's energy escapes into its upper atmosphere (the *corona*) and then into outer space. Sometimes powerful magnetic fields in the surface dam back the flow, and the gas there cools by nearly 2,000°C. The cooler gas—although at 4,000°C—looks dark against the Sun's brilliant surface, and forms a *sunspot*. Spots can reach sizes over ten times that of the Earth.

The Sun is spottier at some times than at others. Every eleven years or so the number of spots builds up to a peak, and the next maximum is due around 1990. Sunspots are just one sign of this ebb and flow of magnetic activity on the Sun. *Prominences*, great arcs of incandescent gas, hang in the corona above the spots, and can unpredictably blast off into space. *Flares*—incredibly violent Sun-storms in the lower corona—also increase in number and intensity near solar maximum, while the strength of the solar wind, a stream of charged atomic particles pouring out of the Sun, increases to gale force.

All these changes affect the Earth. Flares disrupt radio communications, and give rise to spectacular displays of the northern and southern lights near the poles, while the dangerous radiation they produce is a hazard for space voyagers unprotected by Earth's atmosphere. Some scientists believe that the Sun's cycle affects our climate, and even our short-term weather patterns. There's no doubt that we desperately depend on the Sun to be kind.

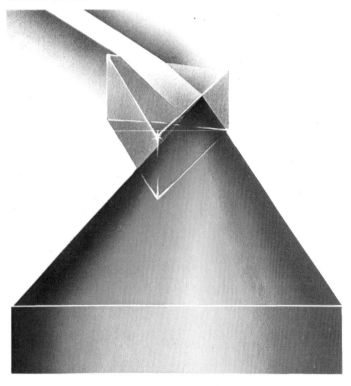

▲ How to make a spectrum, by spreading out sunlight with a prism. Try this experiment with streetlights to see how different their spectra are.

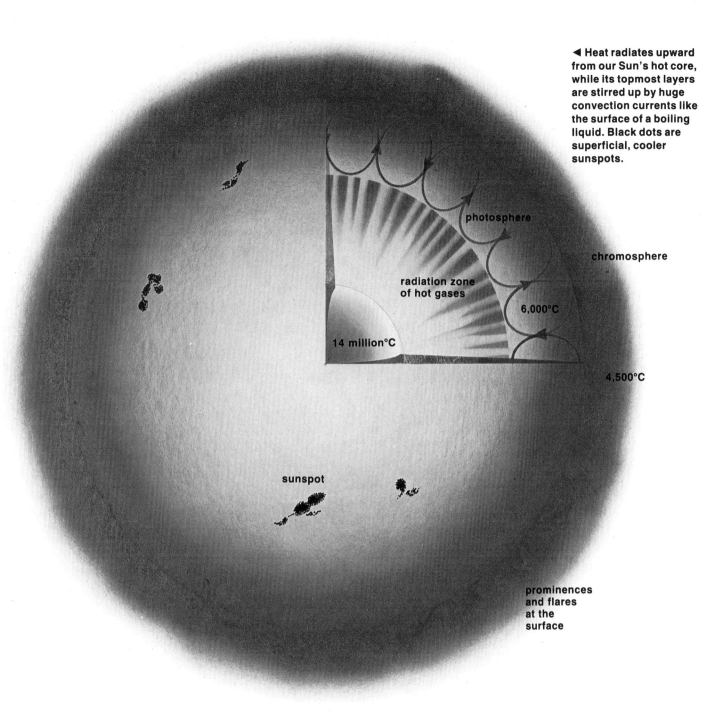

◀ Heat radiates upward from our Sun's hot core, while its topmost layers are stirred up by huge convection currents like the surface of a boiling liquid. Black dots are superficial, cooler sunspots.

Yet anyone seeing the Sun from a few lightyears away would have no inkling of our struggle for existence. All they would see would be a faint, steadily shining star, utterly alone in the blackness of space.

WARNING!
NEVER LOOK AT THE UNDIMMED SUN DIRECTLY—EVEN THROUGH DARK FILM OR TINTED GLASS. AND NEVER, EVER LOOK AT THE SUN THROUGH A TELESCOPE OR BINOCULARS. YOU WILL BE PERMANENTLY BLINDED.

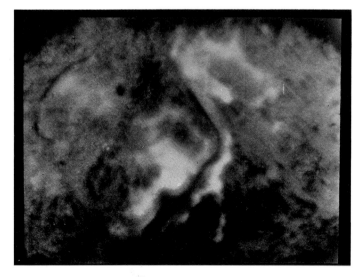

▶ A solar flare like this magnetic Sun-storm can release as much energy as a million H-bombs.

Stars like the Sun

Not even the stars are eternal. Like us, stars are born, they live and they die. But a star's lifespan is so many millions of times longer than ours that an astronomer rarely sees one changing as it ages: and so he has to piece together a star's life story by looking at the variety among the thousands of examples that are around him.

Like our Sun, stars are born in cool clouds of gas and dust which collapse under their own gravity. Contraction makes the hearts of these clouds break up into thousands of globules—*protostars*—which collapse independently to become smaller and hotter. A protostar contracts until its central temperature reaches 10 million degrees. Then the intense heat triggers off nuclear reactions, and a surge of energy floods to the surface and stops the collapse. A young star blazes into being.

Cloaked in mantles of gas and dust, young stars reveal themselves only by the heat (*infrared radiation*) they give out. But a little later they start to light up the cloud around them, and we see it as a fan-like, glowing nebula, like the Orion Nebula. Later still their powerful radiation drives the tatters of the nebula away into space, leaving the young stars loosely clustered together surrounded by just a few shreds of gas and dust. Most of these stars will drift apart, but over half will have formed so close together that they will always be double or multiple stars.

The life of a star

A young star will live for millions of years, but its exact life-expectancy depends upon its mass. A Sun-like star, or one less massive, is assured of living for thousands of millions of years. Stars like these are modest users of their hydrogen fuel, and give out relatively little energy. They shine yellow, orange or red: in star-heat terms, warm, lukewarm and cool—although "cool" is still 3,000°C! Massive stars, on the other hand, shine searing blue-white (over 20,000°C) as they rip through

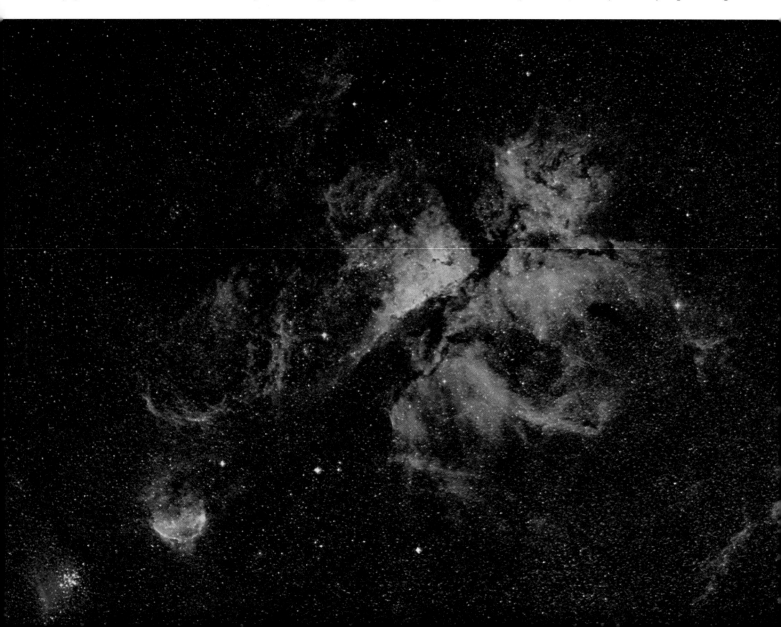

their fuel in only a few million years. But at some time every star's fuel must run out.

For a star like our Sun the end comes after some 10,000 million years. The central hydrogen core is by now all helium: the energy source is dead. Gravity takes over and squeezes the core tight, and at the same time, the outer layers of the star billow out, cooling to dull red. The star has become a vast, tenuous red giant, hundreds of times its previous size. By swelling and shrinking it changes its brightness, and we see several of these long-period variable stars in our skies. But it's all too much for an old star to take. Gently it puffs away its cumbersome envelope into space, which briefly shines as a glowing ring—a planetary nebula—before dispersing.

What's left? A super-collapsed core: most of the matter which made up the original star is squeezed into an object the size of a planet. This white dwarf star is so dense that a teaspoonful of its material would weigh a ton! But it has no energy left. All it can do is to leak away its dwindling heat reserves, growing steadily dimmer until it reaches the stellar graveyard as a cold, black globe.

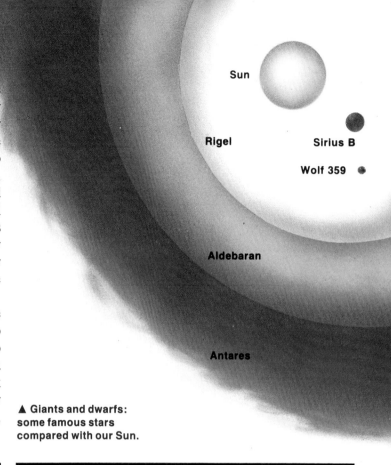

▲ Giants and dwarfs: some famous stars compared with our Sun.

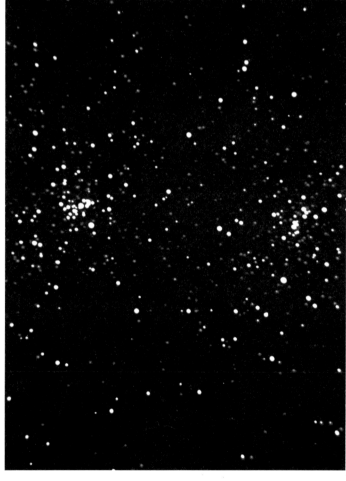

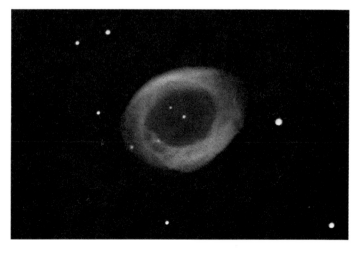

◀ The Carina Nebula hides one of the youngest and heaviest stars known.

▲ The double cluster in Perseus, born only a few million years ago.

▲ Planetary nebula: the Ring Nebula in Lyra, a dying star.

▲▲ Part of a huge gas cloud, the Orion Nebula is a birthplace of stars.

Star death

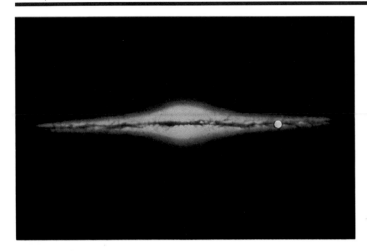

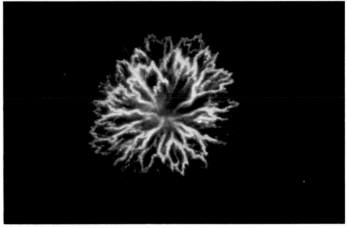

◀ A supernova may be as bright as its galaxy.

▲ A shell of gas ejected in a nova explosion.

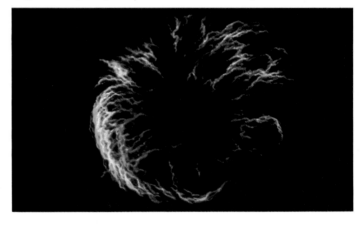

▲ The expanding shell of debris from a supernova explosion can give rise to new stars.

"It appeared like half a bamboo mat in the sky." This was the reaction of a Chinese astronomer to the guest star of AD 1054: a star which lit both night and day, and stayed visible for nearly two years before fading from sight. Today we know that the Chinese astronomers witnessed the gory end of a massive star in a *supernova* explosion; a star whose tangled wreck we now see as the Crab Nebula.

Supernovae are extremely rare. The last to be seen in our Galaxy occurred in 1604, nearly 400 years ago. Less brilliant novae appear much more frequently, but these are tame in comparison and have nothing to do with the deaths of stars.

A star is doomed to die as a supernova if it weighs more than ten times as much as a star like the Sun. Heavyweight stars like these can keep going after their central hydrogen has been converted into helium. They have enough mass to squeeze the helium into carbon, and even when all the helium has been exhausted, the cores can shrink still further to compress the carbon into heavier and heavier elements. All these reactions keep the star shining—until the core comes to be made of iron. When the star tries to fuse iron, a dramatic collapse occurs: iron fusion takes *in* energy rather than giving it out, and the star tries to get this energy by contracting. It's a futile bid, because the extra heat disintegrates the core and the whole star blows up, shining thousands of millions of times more brightly than an ordinary star as it spews its insides into space. But its death is not in vain. Its processed material will, in time, give rise to new stars and planets.

Sometimes a massive star's core survives the blast. If it does, it ends up as something far more bizarre than a white dwarf: a *neutron star*, or even a *black hole*. A neutron star is so compressed that a pinhead of its matter would weigh a million tons; and the entire star measures only 25 km (15 miles) across! Newly formed neutron stars spin several times a second, emitting pulses of energy at astonishingly regular intervals like a cosmic lighthouse beam. Not surprisingly, one of these *pulsars* lives at the heart of the Crab Nebula.

Black holes

If a really massive core is left behind, compression will turn it into gravity's ultimate freak—a black hole. The core shrinks until its escape velocity approaches the speed of light; and then it blacks out. But nothing can stop it from collapsing further. Inside the hole, astronomers believe that it shrinks until it's an unimaginably small point. Anything that comes close enough to fall into the hole—whether a star or a spaceman—will share the same fate. But we can never see their sticky end. Once they're inside the hole, no signal can get out, for nothing travels faster than light. Nor can a victim escape by falling through to another universe as some scientists once thought. A black hole is nature's surest trap.

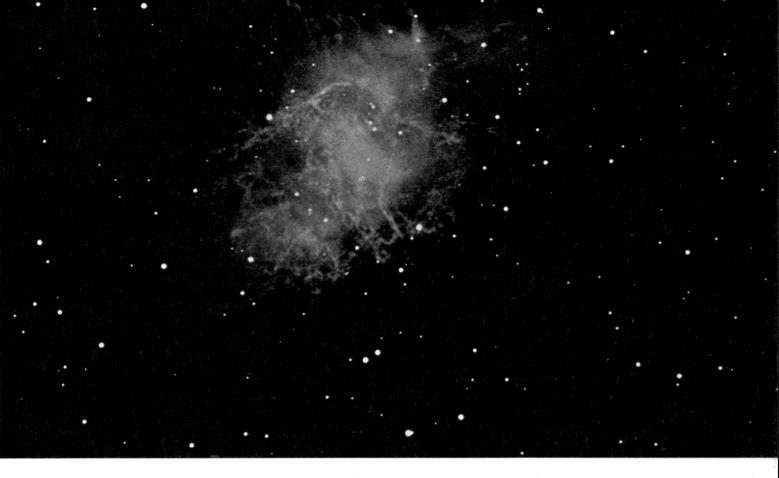

▲ The Crab Nebula—wreck of a star which exploded in AD 1054.

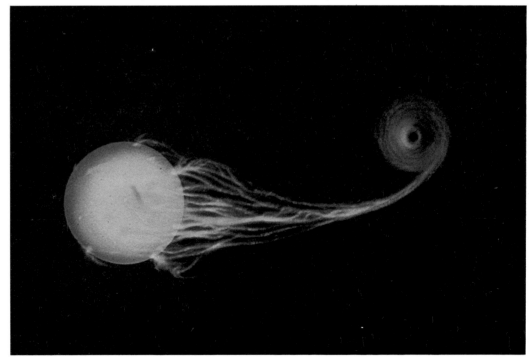

▶ A black hole orbiting a close companion star pares off its outer layers making a whirling vortex of gas around the hole. Astronomers believe that they have now picked up powerful radiation from these *accretion discs*.

5: The scale of the Universe

Let's now travel into distant space to discover the Universe beyond our Solar System

Astronomers need to know the distances to the objects they observe, but this gets harder to measure the further away an object is. A planet's distance can be measured extremely accurately by bouncing a radar beam off it, but you can't get echoes like this from the stars. Even the nearest star is over four lightyears away—so far that its light, moving at 300,000 km/s (186,000 mps) takes over four years to reach us. Here astronomers must use less direct methods, such as watching the tiny shifts in a star's apparent position as Earth moves around the Sun (*parallax*), or looking at the real motions of the members of a star cluster. Galaxy distances are estimated from the brightnesses of stars, although the most remote galaxies are too far even for this.

On these pages we take nine steps to travel from the Earth–Moon system to the limits of the Universe—an increase of 10,000 million million times!

1 Region of Earth-Moon system, 10 million km across (enlarged to show details).

2 The innermost part of the Solar System, out to the orbit of Jupiter (1,000 million km).

3 The Sun and its family all the way out to Pluto (100,000 km).

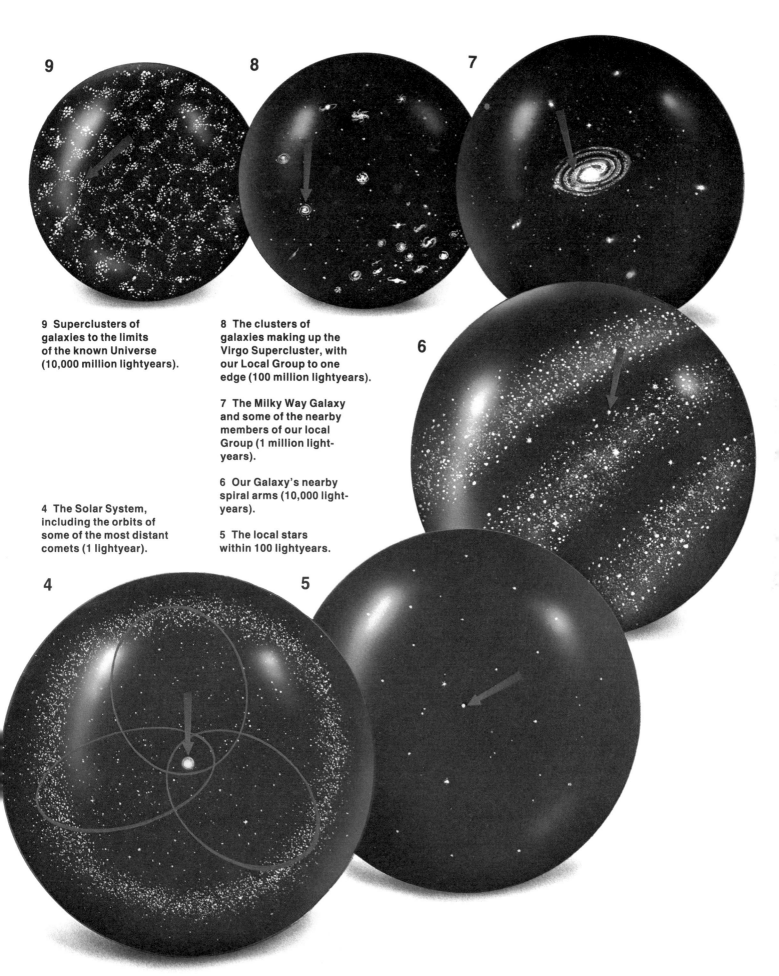

9 Superclusters of galaxies to the limits of the known Universe (10,000 million lightyears).

8 The clusters of galaxies making up the Virgo Supercluster, with our Local Group to one edge (100 million lightyears).

7 The Milky Way Galaxy and some of the nearby members of our local Group (1 million lightyears).

6 Our Galaxy's nearby spiral arms (10,000 lightyears).

5 The local stars within 100 lightyears.

4 The Solar System, including the orbits of some of the most distant comets (1 lightyear).

Galaxies

Imagine that the stars went on forever through space. Everywhere you looked, you'd see a star, and the sky would be ablaze. Somewhere, though, the stars must run out, because the sky between the stars is black.

But look closer. If it's very dark, you should spot the Milky Way, a misty band of light threading across the sky. Sweep this band with binoculars and you will find that it's made up of millions of stars. This means that the stars do go on in one direction: along this narrow strip. Along the Milky Way you look straight into our flattened local star-system—the Galaxy. Every star we see in the sky belongs to it.

Our Galaxy

Seen from the side, our Galaxy would resemble a couple of fried eggs stuck together back to back. Our Sun lives well out in the white; just one insignificant star among 100,000 million. A spacebird's-eye view from above would reveal a vast spiral wheel 100,000 lightyears across, its curving arms of stars, dust and gas revolving slowly about a dense hub of old, red stars. The arms are zones where stars are still forming, triggered into being by supernova shocks and by the Galaxy's own spin. Surrounding this island is a huge spherical halo of old stars and clusters, a relic of the days when our Galaxy first formed.

▲ The small spiral galaxy M 33 contains about 10,000 million stars.

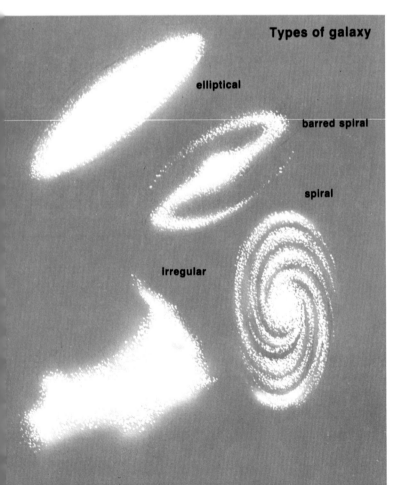

Types of galaxy

elliptical

barred spiral

spiral

irregular

Beyond the Galaxy—as our friends from OVOG-1 found—space is empty. But here and there we find other galaxies: and space is so vast that there must be over 1,000 million of them. You can even see the nearest in the sky without a telescope. The Andromeda galaxy (alias OVOG-1) can be seen as a misty patch in autumn skies, while those living in the southern hemisphere can see two even closer galaxies, the Large and Small Magellanic Clouds.

Spirals, ellipticals and irregulars

Even in a small telescope, galaxy numbers run into hundreds. About half are spiral galaxies, like our Milky Way and the Andromeda galaxy. Astronomers who love galaxies find their different spiral shapes as individual as the different members of a large family of puppies. All of them are alike, though, in having a central *nucleus* surrounded by a flat disc rich in the raw materials from which stars are made. Similar, too are the few *irregular* galaxies (like the Magellanic Clouds), which are busily forming stars but seem too small to grow spiral arms.

The other main type of galaxies are ellipticals. These galaxies are like the hubs of spirals stripped of their

▲ Seen almost edge-on, spiral galaxy NGC 253 shows a thick band of obscuring dust in its disc.

▼ An elliptical galaxy made up of old, red stars.

▲ A galaxy like our own: the Andromeda galaxy.

arms: huge balls of old, red stars, puzzlingly swept clean of the dust and gas necessary to form new generations. Ellipticals range widely in size. Some of them contain a million million stars, making them the biggest galaxies in the Universe. Others are so tiny that they have fewer stars than a globular cluster, and they're so sparse that an unfortunate inhabitant would only ever see about five stars in the night sky!

Galaxy swarms

Galaxies are gregarious and like to live in clusters or groups. Our own Local Group has nearly 50 members, including our Galaxy and the Andromeda galaxy, all arranged in a flattened cluster about five million lightyears across. But it pales into insignificance against giants like the Virgo or Coma clusters, whose thousands of galaxies swarm across regions over 20 million lightyears in diameter. And finally there are super-clusters—clusters of clusters of galaxies—which at 150 million lightyears across, are the biggest clumpings of matter in the Universe.

▶ The nearest galaxy to ours is this irregular galaxy called the Large Magellanic Cloud.

The violent Universe

Over the past twenty years it has become clear that some galaxies are not the peaceful collections of stars we once thought they were. Out of every hundred galaxies, two or three are exploding. Even our own Galaxy shows the traces of a disturbance at its heart.

Exploding galaxies

Generally, though, it is only the largest galaxies which suffer in this way. Many giant ellipticals have disturbed hearts, and some even have jets of matter thousands of lightyears long spewing out of their cores. Some, called *radio galaxies*, are surrounded by huge clouds of gas which emit radio waves, ejected because of disturbances in the galaxy. Most violent of all are the quasistellar objects, or *quasars*: galaxies so remote that they appear as no more than starlike points of light.

In all cases astronomers believe that the source of the disturbance is a vast black hole—perhaps a thousand million times more massive than a star—which lurks at the galaxy's heart. Only huge galaxies can accommodate a black hole this big, believed to have formed early on in a galaxy's history when it was a rapidly collapsing gas cloud. The hole is surrounded by a glowing vortex of infalling, whirling matter—an *accretion disc*—made of gas and ripped-apart stars. It is fiercely hot; it spins at breakneck speed; and it is a potent source of energy to generate the awesome power of an active galaxy. Exploding galaxies tell us a great deal about how galaxies form and evolve.

The Big Bang theory

Whether exploding or peacefully whirling, all galaxies have one thing in common: they are all rushing away from one another. Like the skin of a balloon, space between the clusters of galaxies is stretching as our Universe expands, and we can follow this outward flight by analyzing the spectrum of light from remote galaxies.

If we backtrack the motion, we find that all the matter in our Universe was together some 15,000 million years in the past. It occupied a single point, and its density must have been unimaginably high. We don't know how long it sat there: it is meaningless to speculate, as time had no meaning before events began to happen. The first event of all was the *Big Bang*. The point exploded, and our Universe was born.

The young expanding Universe was filled with hot gas and radiation, which sensitive radio telescopes pick up even today as a faint background hiss. But not until millions of years after was it cool enough for thin clouds of gas to swirl and tumble and collapse to form the

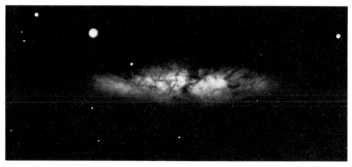

▲ Galaxy M 82 spews gas streams out into space.

▼ A cluster made up of thousands of galaxies.

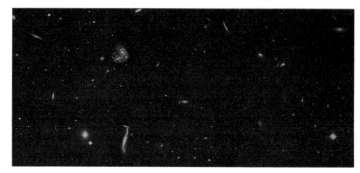

▲ A conventional photo of M 87. The jet of gas may come from the region of a huge black hole.

galaxies. And 10,000 million years after that cataclysmic birth, a small star we call the Sun quietly came into being in an obscure part of an anonymous galaxy.

The future of the Universe

From its past performance we can try to forecast the future of the Universe. If there was sufficient matter in the Universe, its gravity could halt the outward expansion: but it doesn't seem as if there is enough. At least this means that our remote descendants will not have to dread a Big Crunch as the Universe recollapses. At the moment it seems that space is destined to become progressively emptier. The far future will be lonelier and colder than we could ever imagine.

▶ Heart of an exploding galaxy: a dramatic shot of M 87, revealing a jet of matter thousands of lightyears long issuing from its core.

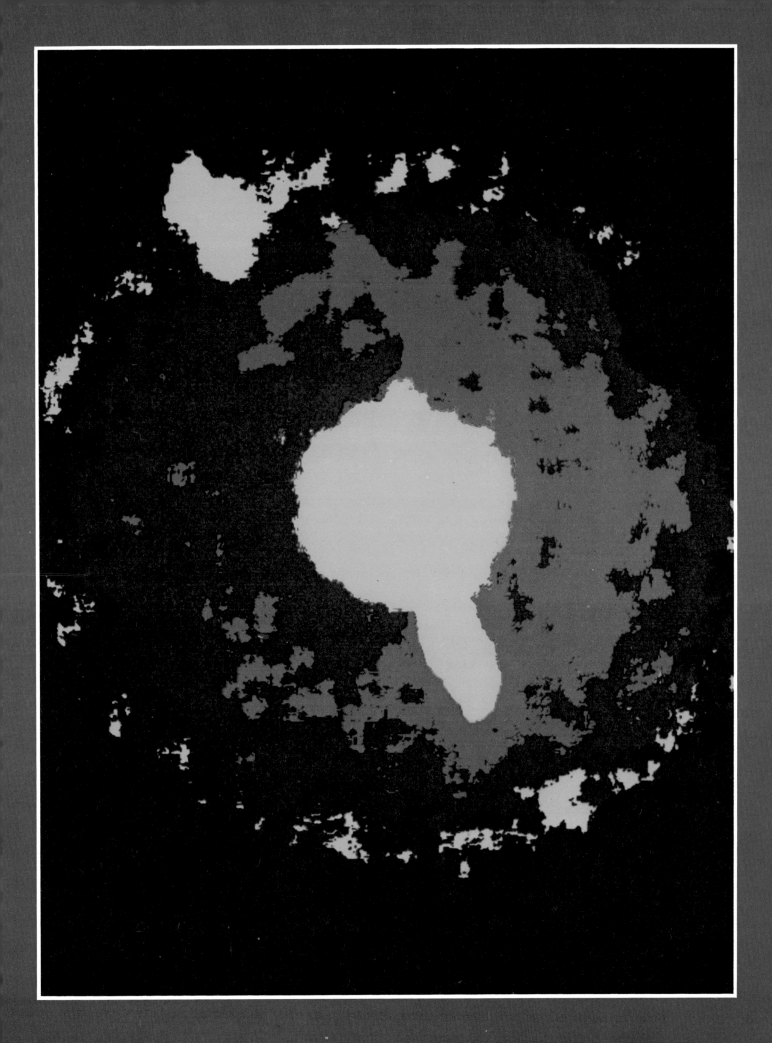

6: How astronomers work

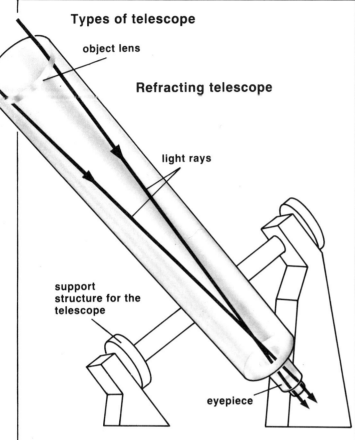

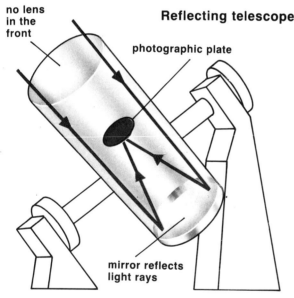

▲ How telescopes work: refractors collect light with a lens; reflectors collect light by means of a mirror. The world's biggest telescopes today are all reflectors.

6.0 m Zelenchukskaya, USSR
5.1 m Mt Palomar, USA
4.5 m MMT, Arizona, USA
4.0 m Kitt Peak, USA
4.0 m Cerro Tololo, Chile

3.9 m Siding Springs, Australia
3.8 m Mauna Kea, Hawaii
3.6 m La Silla, Chile
3.0 m Mt Hamilton, USA
2.7 m Fort Davis, USA

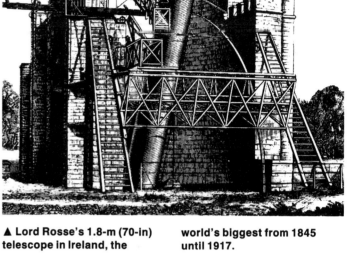

▲ Lord Rosse's 1.8-m (70-in) telescope in Ireland, the world's biggest from 1845 until 1917.

Telescopes are not the only means of discovering the Universe. Radio waves, X-rays and infrared rays are being used to open up the skies.

Cartoons always depict astronomers as white-bearded old men peering shortsightedly through rickety telescopes. Nothing could be further from the truth. Today's professional astronomers are often surprisingly young; many are women; and none would ever be seen dead squinting through a telescope! That's because there are much more efficient means of recording the elusive light from the remote objects in space than the unreliable human eye.

But the telescope is still the astronomer's main tool. Although it has developed considerably since Galileo's "optick tube" of 1609, it still performs the same task: to collect more light than the eye, so that an astronomer can learn more about faint objects.

The first telescopes were of two types. Galileo's kind had a large lens at the far end to collect light (the *object glass*), while the astronomer looked through a small lens (the *eyepiece*) to magnify what he saw. Telescopes like these are called *refractors*. Isaac Newton was bothered by the rainbow which is produced by refractors (the lenses act like prisms to produce a spectrum), and so he developed a *reflector*, which collected light with a large curved mirror like a shaving mirror. The world's biggest telescopes today are all reflectors.

Until a century ago astronomers had only one way of

▲ The Hale 5-m (200-in) telescope, the world's largest until it was overtaken by the Russian's 6 m (236 in) in 1976. It is used by astronomers from all over the world.

▼ A new concept in cheap telescope design: the Multiple Mirror Telescope in Arizona has six 1.8-m (70-in) mirrors giving the total power of a 4.5-m (180-in) instrument.

recording what they could see through their telescopes, and that was by looking. Observing was a long, chilly business.

Astronomers would stay at the eyepiece for hours, waiting for rare periods when the air was still before hurrying, in that instant, to make their measurements or record elusive details.

Photography changed everything. By attaching a photographic plate to a telescope in place of his eye, an astronomer was able to record much more detail than he could ever hope to see, particularly if he made a long exposure.

And so the twentieth-century generation of telescopes are mostly giant space-cameras, with big mirrors to collect as much light as possible and accurate drives to follow an object as the Earth revolves during long exposures.

But astronomy is not simply a matter of taking picture-postcard views of the Universe. Astronomers are concerned with analyzing the light picked up by their telescopes: dissecting, chopping, stacking, filtering and examining it to find out just *how* it was produced in one particular object lightyears away. We have already learned that starlight spread out into a spectrum can tell of conditions there: but to record incredibly faint spectra, a modern astronomer will use a battery of devices, including TV cameras, which would seem rather more at home in a film studio than an observatory. The results usually come out as arrays of numbers, which the astronomer feeds straight into a computer. A full analysis can take months—even years.

Invisible astronomy

radio waves **radar waves** **microwaves** **infrared rays**

Imagine how difficult it would be to compose a tune using just the two middle notes on a piano! By concentrating exclusively on the light given out by bodies in space, astronomers themselves were in a similar situation until quite recently.

Light is a vibration called an *electromagnetic wave*. Each train of lightwaves has a certain *wavelength* (the distance from the peak of one wave to the next), and our eyes perceive different wavelengths as different shades. Red light has a longer wavelength than yellow; yellow waves are longer than blue. But when you consider that the wavelength of red light is 700 nanometres (700 millionths of 1 mm!) and that of blue is 400 nanometres, the difference is minute.

▼ **The Parkes radio telescope in Australia** gathers radio waves in a dish 64 m (210 ft) wide.

visible light   ultraviolet light   X-rays   gamma-rays

▲ Lightwaves occupy only a tiny part of the electromagnetic spectrum which stretches from ultra-short gamma-rays up to long radio waves.

Although our eyes are particularly sensitive to this narrow band of vibrations, light pales into insignificance when compared to the whole range of electromagnetic waves. These range from ultra-short gamma- and X-rays with wavelengths of a million millionth of 1 mm, through ultraviolet, light and infrared rays, up to radio waves as much as 20,000 m long. Bodies in space emit radiation over the whole of this *spectrum*, and it's only recently that we've been able to tune into any of the other bands!

Radio telescopes

The radio band was the first to be probed because radio waves, like light, penetrate our atmosphere. The first radio waves from space were picked up accidentally by Carl Jansky in 1931, but radio astronomy really took off after World War II. Radio telescopes work just like optical reflecting telescopes, but to see the same amount of detail, the reception dishes must be much larger to accommodate the faint, long-wavelength radio waves.

Despite increasing terrestrial interference (even from microwave ovens!) radio telescopes—especially when linked up by clever electronic techniques to synthesize dishes many km across—can do as well as optical telescopes. They can see through the opaque dust clouds in space which dog optical astronomers, and trace objects—such as ultra-hot gas clouds—which don't give out light. Radio galaxies, pulsars and quasars were all discovered by radio telescopes.

Infrared telescopes

Infrared rays are heat rays, and they tell astronomers of darkly cocooned places where young stars are forming. Ironically, infrared telescopes (like the new British 3.8 m in Hawaii) have to be placed on the coldest, highest mountains of the world because the rays would otherwise be absorbed by the water content in our atmosphere!

It's fortunate that our atmosphere does absorb those

▲ Einstein orbiting observatory photo of X-rays from quasar 3C 273.

wavelengths shorter than light. The dangerous ultraviolet, X- and gamma-rays would otherwise turn our planet into a sterile wasteland and destroy all life. But by sending up rockets and satellites into space to detect these rogue rays, X-ray astronomers learn of exploding stars, violent galaxies and—possibly—black holes.

Space telescopes

The future of optical astronomy, too, lies in space. Hopes are pinned on a 2.4-m telescope due to be launched into space by the U.S. Space Shuttle early in the 1980s. As well as pinning down distances ten times more accurately, the space telescope will be able to see faint objects 100 times greater than any previous telescope.

Astronomy's current revolution focuses on the invisible radiations. But who can tell what's in store when optical telescopes at last break through our gray, churning atmosphere and soar into space?

7: Is anyone there?

As yet there is no firm evidence of life beyond Earth, but messages from our planet are being sent into space.

As our fictional friends from OVOG-1 approached the Solar System (see Chapter 1), I wonder whether any of their instruments detected any other spacecraft.

Presently two spacecraft, Pioneer 10 and 11, are making their way beyond the Solar System to wander forever through our Galaxy. Soon they will be joined by two Voyager craft also destined to drift forever between the stars.

It has been estimated that the Pioneers will be as distant as the nearest star in about 80,000 years' time.

If the Voyager missions proceed as planned, it is likely that both craft will pass within one lightyear of a red dwarf star in about 60,000 years time. Presently situated in the constellation of the Little Bear, AC + 793888, as it is called, is a small cool star much older and redder than the Sun. Perhaps there are planets surrounding AC + 793888, and perhaps in 60,000 years' time there could be intelligent creatures there who could retrieve and examine our two Voyager ambassadors.

By that time the Pioneers and Voyagers will be showing the scars of thousands of years of interstellar travel. Pock-marked, battered and worn by meteorites, each will still carry messages from the planet of their origin.

Both Pioneer craft will carry plaques giving information. There is a diagram of the Solar System showing the paths of spacecraft through the planets. It demonstrates how to locate Earth with respect to some nearby pulsars, and through the coded information given, pinpoints the particular period of time in which the craft were built and dispatched.

Also etched on the plaque are two human beings—one male and one female. The male has his arm raised in what, it is hoped, will be understood as a gesture of peace.

The Voyager craft also carry such plaques. In addition they carry a gold-plated phonograph record and a special player. The records contain, encoded in the audio spectrum, 117 pictures explaining Earth and Man, greetings in 52 languages, and a selection of sounds of the Earth. These sounds include those of the creatures of land, sea and air, the sounds of rain, thunder and waves, the sounds of rockets, trains, cars, and 90 minutes of some of the world's best music.

The chance of these plaques and records ever reaching extraterrestrials is certainly very remote. However, as a gesture, their inclusion in the spacecraft was worthwhile if only to show that at one time a species on Earth was capable of building a craft destined to reach the stars.

UFOs

A few hundred years ago it was possible to be burned at the stake for saying that life could exist on other worlds. Today all over the world a surprising number of people believe that other beings exist and live on worlds far beyond our own. There are a few who believe that our own planet has been visited by extraterrestrials in the past. There are also those who believe that UFOs are spacecraft piloted by creatures from distant stars.

However, it must be said that at present there is no evidence at all to suggest the existence of life on any other body apart from Earth.

Of course, thousands of people have seen numerous lights, glows and moving bodies in the sky—and so they should. At any one time there are thousands of objects in the orbit around the Earth. The atmosphere is filled

▲ Although many people have reported seeing UFOs, there's no evidence to suggest that they are alien spacecraft of the kind pictured here.

▲ Our fastest spacecraft, Voyager, pockmarked and worn after years of interstellar travel.

▶ If aliens do intercept Voyager, the symbols on its record canister will tell where it came from.

with aircraft, civil and military, and numerous objects such as research balloons—so there's plenty to be seen.

It may sound surprising, but almost 90 per cent of all sightings reported can usually be accounted for by two of the planets, namely Venus and Jupiter. These two objects have often been photographed, followed and even feared by anxious UFO spotters.

In addition to planets, satellites, aircraft and balloons, don't forget there are meteors, fireballs and ball lightning.

Usually the vast majority, perhaps as high as 98 per cent, of UFO sightings are termed MFOs—Misidentified Flying Objects. Very often no further information can be obtained on the other sightings as the reports are too vague—a ball of light, a glowing disc moving slowly or a bright ball with sparks.

As yet there is not one piece of really good evidence to support the theory that UFOs are spacecraft from other worlds. There is even less evidence to show that any have landed on Earth either in the past or at the present time.

Communicating with aliens

Apart from plaques on spacecraft, which are very much a hit and miss affair, we have a much more precise means of sending messages to the stars.

The giant radio telescopes built by astronomers to explore the radio universe can also be used to transmit radio signals. In fact the huge telescope at Arecibo, Puerto Rico, is so powerful that it could communicate with a similar telescope situated anywhere within the Milky Way Galaxy.

The problem of radio communication with distant civilizations is time. Radio waves travel at the same speed as light. The nearest star is four lightyears away, so if a civilization existed there, and if it could receive our messages, a conversation would be a very long, drawn-out affair. It would take four years for our message to reach them and another four years for their reply to reach us. There are good reasons to believe that there is not a civilization so close, so the time lag gets worse. With greater distances there is the possibility that the person who sent the message will be long dead before a reply is even halfway here!

Messages from Earth

Spreading out from our Solar System at the speed of light is a shell of radio noise now about 80 lightyears across. This noise is composed of radio, television and other transmissions intended for use on Earth, but which have leaked out into space. Perhaps some distant and puzzled astronomers are trying to decipher a rather faint transmission of "The Lone Ranger"!

Deliberate attempts have been made to listen for signals sent to us from other worlds. As yet nothing has been detected. We have also sent radio messages to various stars. The most interesting and most controversial was the message transmitted from Arecibo in 1974. This signal, boosted by the giant radio telescope, made our planet on one specific wavelength a transmitter 1,000 times more powerful than any other object in the Galaxy.

The telescope beamed its message in the direction of M 13, a huge collection of old stars about 50,000 lightyears away from us. If it is received, understood and answered, it will be 100,000 years before a reply arrives on Earth. Will anyone be here to receive it or even remember it was sent?

But will any alien be able to understand our message? Just how do you communicate with extraterrestrials about whom we know absolutely nothing? After much planning and guessing, it was decided to send a message in a very simple code.

A series of two different tones was sent. One tone

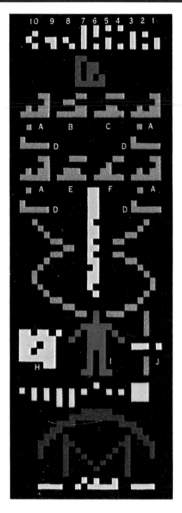

▲ This radio message beamed to the stars tells of our existence.

▲ Would an alien spacecraft look like this? Your guess is as good as any!

represented a dark spot, the other a light spot. When arranged in a particular way in the form of a rectangle, a grid picture is formed.

This picture tries to show what we look like. It also includes a drawing of the DNA molecule, the building block of life. It also shows our own planet, the telescope, our position in the Solar System, and much more information.

Some astronomers feel that we should not transmit such messages to the stars, for they argue there is no way of telling who or what will reply. There is no guarantee that the recipients of our message will be friendly. For the same reason it is argued we should not reply to any messages received from the stars. But is there anyone there for us to contact?

Today a majority of astronomers would agree that statistically the chances of life existing elsewhere are reasonable. Already we know that the basic components necessary for the building blocks of life exist

throughout space. What we don't know is whether the necessary conditions for the evolution of intelligent lifeforms exist elsewhere.

Extraterrestrial life

Astronomer Frank Drake proposed a formula to give the number of civilizations in our Galaxy. Drake's formula takes into account the rate of star formation in the Milky Way; the fraction of those stars that develop planets; the number of such planets ecologically suited for life; the percentage which evolve to intelligence; the fraction of those which can communicate over interstellar distances; and the lifetime of such a civilization.

Dependent upon the figures entered to represent each variable, we come up with different answers. A pessimistic answer is between 40 and 40,000 advanced civilizations. The most optimistic is that as many as one million technologically advanced civilizations could exist in our Galaxy.

Not all astronomers would agree with these more optimistic figures. Some say that the figures do not take into account the special factors which have helped man evolve here on Earth and which are probably lacking elsewhere.

Let's hope the pessimists are wrong. There are so many galaxies, so many stars and perhaps so many planets in the known Universe that it would appear that there are possibilities for all kinds of life to evolve.

What the future holds

While the difficulties in contacting distant civilizations are immense, and our own resources so small, the passage of time can change all these things.

In the lifetime of many people on Earth today, vast changes have taken place. It is not so long since flight was a dream, computers unheard of, television and radio unthinkable. Remember it is only 24 years since the English Astronomer Royal stated that the idea of spaceflight was pure bunkum!

If there is anyone there, and we ever manage to make contact, think of the wonderful possibilities, the fascination of meeting new minds and exploring new worlds. It could be the most exciting discovery in the history of Man.

Watching the skies

Binoculars

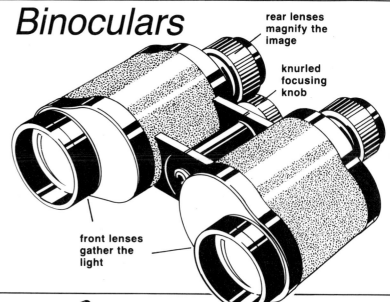

The fun about astronomy is that it is a science which most people can appreciate and enjoy. Although the days are gone when amateurs could make great contributions to research astronomy, there is an immense amount of space exploration which can be conducted from the surface of our own spaceship Earth.

With our eyes alone we can explore the stars, learn the star patterns, enjoy meteors and stand in awe of eclipses and the eerie beauty of the Aurora.

We can make our eyes work better for us by collecting more light from the stars and planets. This is done by using binoculars, telescopes and cameras.

Telescopes

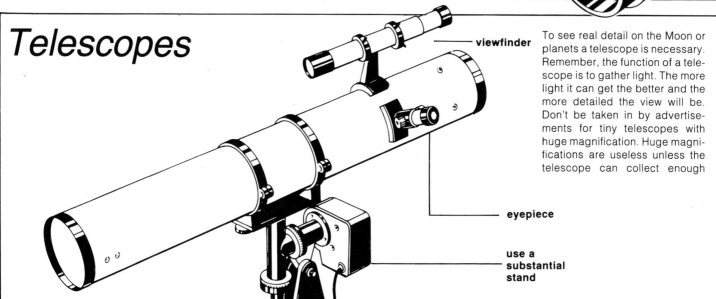

To see real detail on the Moon or planets a telescope is necessary. Remember, the function of a telescope is to gather light. The more light it can get the better and the more detailed the view will be. Don't be taken in by advertisements for tiny telescopes with huge magnification. Huge magnifications are useless unless the telescope can collect enough

Taking photographs

Any camera which has a B or brief setting in the shutter can be used to photograph the stars. If placed on a steady base, pointed at the sky and given an exposure of a few minutes, the stars will be seen. Longer exposures will show the stars as streaks; shorter exposures will give smaller streaks or points of light. Experiment with film speeds of 200 or 400. ASA colour film reveals the different shades of the stars quite well. If the camera has a variable aperture, it should be set at its widest setting. Be sure to tell the film processors that you have photographed stars. They could think your film was blank and not print it!

A simple camera can be used to photograph the Moon through binoculars or a small telescope.

For beginners a pair of binoculars and a star map are all that are required to begin sky exploration. Binoculars come in various types and magnifications. They are usually referred to as 7 × 40's, 8 × 30's, 10 × 50's, etc. The first, usually smaller, figure gives the magnification, and the second, usually much larger, is the diameter in mm of the main lens of the binoculars. It is best to buy a pair of binoculars which are not too heavy—remember you'll be wanting to look high in the sky. Try to get a pair with large lenses at least 50 mm in diameter. Don't get too high a magnification—10 × is high enough. Any higher and you'll

find them impossible to hold steady without some sort of tripod or support. A pair of 10 × 50's is a good buy. This means they have lens diameter of 50 mm and a magnification of 10 times.

With binoculars the sky begins to open up. The Moon appears cratered and scarred. Jupiter's satellites zoom around their parent star, and the Milky Way resolves into clouds of stars. Stars of every hue appear scattered across the sky like gems. Many comets are seen beautifully in binoculars, and even bright satellites can be followed.

Remember to wrap up well:
two pullovers
two pairs of socks
warm hat
thick socks or wellington boots
coat or anorak
gloves

light. Using high magnification on a small telescope is like using a magnifying lens on a television screen—the picture will get worse, not better. A top magnification of about 40 × per 25 mm of collecting lens or mirror is a good guide.

Astronomical telescopes come in two basic types—refractors and reflectors. A refractor is a telescope which uses a lens to gather light which is brought into focus at the end of a long tube. At the end of the tube a smaller microscope called an eyepiece is used to examine the focused image in detail. To see good detail of the planets a telescope with a 75 mm main lens is ideal.

Smaller than this, less detail is visible. Don't buy small telescopes with lenses between 25 and 40 mm in size. Either buy a pair of binoculars or keep the money and save for a larger telescope. This is better than being disappointed by a tiny telescope incapable of showing any worthwhile detail.

A reflector is a telescope which uses a mirror to collect the light. Reflectors come in a variety of designs and shapes. The smallest reflector to try for is one with a mirror 150 mm (6 in) in diameter.

A 150 mm reflector or 75 mm refractor will enable the rings of Saturn, phases of Venus, clouds of Jupiter, and polar caps on Mars to be seen.

Both types of telescope are fairly expensive when new, but can often be bought secondhand. It is always worthwhile to enquire from your local astronomical society about secondhand telescopes.

Many amateurs prefer to make their own telescopes. Reflectors are much easier to make than refractors, and the saving in cost can be quite substantial. Some people even prepare their own mirrors, but many more purchase prepared mirrors and parts of a telescope to assemble like a kit. If you haven't much money, this is a good way to get a large telescope and spread the cost.

Remember that any telescope requires a good tripod or mount to keep it steady. A telescope also needs to move in different directions. It has to track the stars and planets as they drift across the sky. The greater the magnification on the telescope the shorter the time the object will stay in view.

Larger and more expensive telescopes come equipped with electric motors and drives which enable them to follow objects and track them for long periods of time. Accurate tracking is necessary if you plan to photograph stars and planets through your telescope.

Simply focus the telescope for your eyes. Set the camera at infinity. A speed of 1/30 of a second at 100 ASA will give good views of the Moon. Experiment with different exposures and don't forget to try the instant cameras, some of which give excellent results.

Single lens reflex cameras (SLR) which enable the photographer to see through the camera lens are very useful in telescopic photography. With such a combination the photographer is able to see the telescopic view of the planet or star to be photographed.

It is possible to buy adaptors which allow most cameras to be coupled to a telescope. Good photographs require a lot of experimentation, good records and considerable patience.

If you haven't a telescope, binoculars or a camera, don't forget that you can still explore and enjoy the sky. Clubs and societies often lend telescopes to members, and most astronomers will allow other interested people to look through their instruments.

At public observatories and planetariums it is often possible to view the Moon and planets, even to take photographs through a large telescope, so why not contact the one near you.

Happy skywatching!

Try taking a photo of the Pole Star with a long exposure— you should get a shot rather like this.

Going further

If you have enjoyed reading *Heavens Above* and want to learn more about astronomy and space, there are plenty of directions you can take. Your local bookshop or library can help with books. Nearby there might be an observatory or a planetarium which you can visit. And if you really want to make a hobby out of exploring the night sky, you will find that most large towns have an astronomical society (ask for the address at your public library). Don't be afraid to join: most societies have extremely active under-18 sections who usually outwit the older members on quiz nights!

Here are lists of basic books and places to visit to get you off to a good start.

Reading
Couper, Heather. *Exploring Space*. London: Marks & Spencer, 1980
Henbest, Nigel. *Spotter's Guide to the Night Sky*. London: Usborne, 1979
Man, John (ed.). *Encyclopedia of Space Travel and Astronomy*. London: Octopus, 1979
Moore, Patrick. *The Amateur Astronomer*. London: Lutterworth, 1978
Moore, Patrick. *Yearbook of Astronomy*. London: Sidgwick & Jackson, annually
Ridpath, Ian. *Stars and Planets*. London: Hamlyn, 1979

For readers in USA and Canada
Asimov, Isaac. *The Collapsing Universe: The Story of Black Holes*. New York: Walker & Company, 1977
Asimov, Isaac. *To the Ends of the Universe*. New York: Walker & Company, 1976
Berger, Melvin. *Planets, Stars, and Galaxies*. New York: G.P. Putnam's Sons, 1978
Freeman, Mae and Ira. *The Sun, the Moon, and the Stars*. rev. ed. New York: Random House, 1979
Gribbin, John. *Astronomy for the Amateur*. New York: McKay, 1977
Mitton, Jacqueline and Simon. *Concise Book of Astronomy*. Englewood Cliffs, N.J.: Prentice-Hall, 1978
Young, Louise B. *Earth's Aura*. New York: Alfred A Knopf, Inc., 1977

▼ Two visitors get a close-up view of the star projector at the Greenwich Planetarium.

Looking
Many observatories and planetariums have good shops where you can buy books, photos and slides. There may also be astronomers to hand who can answer sticky questions!
Armagh Planetarium, College Hill, Armagh, Northern Ireland (0861-523689)
Advanced planetarium (seats 100) which gives shows to schools and the public. Hall of Astronomy has modern space and astronomy displays, with experiments you can do yourself. There's a 41-cm (16-in) reflecting telescope available on certain nights which you can look through or even photograph the night sky.
Jodrell Bank, Nuffield Radio Astronomy Laboratories, Jodrell Bank, Macclesfield, Cheshire (04777-339)
The concourse building has a planetarium, and there are many displays and working models. You can see the radio telescopes, including the famous 250 ft Mk 1a.
London Planetarium, Marylebone Road, London NW1 5LR (01-486 1121)
Biggest planetarium in the UK (seats 420), giving shows for both schools and the public. Laserium displays in the planetarium.
Merseyside County Museums, William Brown Street, Liverpool L3 8EN (051-207 0001)

Astronomy displays and planetarium (seats 67), giving schools and public shows.

National Maritime Museum, Greenwich, London SE10 9NF (01-858 4422)
The main museum buildings house historic astronomical globes and navigation instruments. Up the hill is the **Old Royal Observatory**, whose collections cover 300 years of British astronomy. **The Greenwich Planetarium** is housed here in a former telescope dome (seating 50), and gives shows to schools and to the public. There are astronomy courses for children and teachers, during which participants can use the great 28-in refractor.

Royal Greenwich Observatory, Herstmonceux Castle, Hailsham, Sussex (032-181-3171)
Displays in the castle tell of the research done at this, the UK's leading astronomy institution.

Royal Observatory, Blackford Hill, Edinburgh EH9 3HJ (031-667-3321)
The Visitors' Centre describes the research done here and with the new UK telescope in Australia.

Science Museum, Exhibition Road, London SW7 2DD (01-589 3456)
Astronomy is covered in the Exploration Gallery, Astronomy Gallery, and Stardome (small planetarium).

United States of America and Canada
Adler Planetarium and Astronomical Museum, Chicago, Illinois

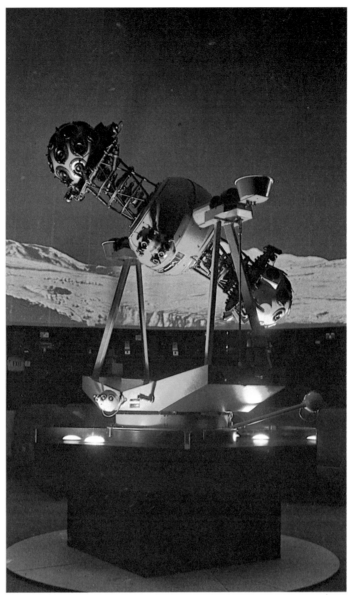

▲ A futuristic moonscape frames this view of the Armagh Planetarium's star projector.

American Museum—Hayden Planetarium, New York, New York
Buffalo Museum of Science, Buffalo, New York
Buhl Planetarium and Institute of Popular Science, Pittsburgh, Pennsylvania
Children's Museum, Nashville, Tennessee
Dominion Government Astrophysical Observatory, Victoria, British Columbia
Flandau Planetarium, Tucson, Arizona
Franklin Institute, Philadelphia, Pennsylvania
Griffith Observatory and Planetarium, Los Angeles, California
Hansen Planetarium, Salt Lake City, Utah
Kansas City Museum of History and Science, Kansas City, Missouri
Kennedy Space Center, Cape Canaveral, Florida
Johnson Space Center, Houston, Texas
McLaughlin Planetarium, Toronto, Ontario
McMaster University Planetarium, Hamilton, Ontario
Minneapolis Public Library, Science Museum and Planetarium, Minneapolis, Minnesota
Montana State University Planetarium, Missoula, Montana
Morrison Planetarium, San Francisco, California
Mount Palomar Observatory, Palomar Mountain, California
Mount Wilson Observatory, Mount Wilson, California
Museum of Science, Boston, Massachusetts
Ohio State University, Department of Astronomy Planetarium, Columbus, Ohio
Ontario Science Center, Toronto, Ontario
The Pacific Science Center, Seattle, Washington
The Planetarium and Science Museum, Vancouver, British Columbia
The Planetarium, Houston, Texas
The Planetarium, Portland, Oregon
Queen Elizabeth Planetarium, Edmonton, Alberta
Smithsonian Air and Space Museum, Washington, D.C.
The Space Center, Huntsville, Alabama
Strasenburgh Planetarium, Rochester, New York
United States Naval Observatory, Washington, D.C.

Index

accretion disc 45, 50
Alcor 33
Aldebaran 32, 34, 35, 43
Algol 32, 34, 35
Alphard 32, 34
Alpheratz 32, 34
Altair 32, 33, 34
Andromeda 32, 34, 35, 48, 49
Antares 32, 33, 43
Apollo 17 18
Aquarius 32, 34
Aquila 32, 34
Arcturus 32, 33
Argo 32, 34
Aries 32, 34
asteroid 23, 29
astrology 39
ATS satellite 18
Auriga 32, 34, 35

Betelgeuse 32, 34, 35
Big Bang theory 12-13, 50
Big Dipper see Plough, The
binoculars 60-1
black hole 7, 44, 45, 50, 51, 55
Boötes 32, 33, 34

Callisto 22, 23
Cancer 32, 34
Canis Major 32, 34, 35
Canis Minor 35
Capella 32, 34, 35
Capricornus 32, 34
Carina Nebula 42
Cassiopeia 32, 34, 35
Castor 32, 34, 35
Cepheus 32, 34
Ceres 29
Cetus 34
Charon 27
Coma 49
comet 28-9, 61
constellation 32-5, 39
Cor Caroli 32, 34
Corona 32, 33, 34
corona 36, 37, 41
Corvus 32
Crab Nebula 44
Cygnus 32, 33, 34

day 30
Deimos 20
Delphinus 32, 33, 34
Deneb 32, 33, 34
Draco 32, 34
Drake, Frank 59

Earth 13, 15, 16, 17, 18-19, 21, 22, 28, 30-31, 37, 46
eclipse 36-7
ecliptic band 38-9
electromagnetic wave 54
Epsilon Pegasi 32
Eridanus 32, 34
Europa 23

fireball 28, 29, 57
Fomalhaut 34

galaxy 12, 44, 46, 47, 48-51, 55, 58, 59
Galileo, Galilei 16, 52
gamma-ray 52, 55
Ganymede 23
Gemini 32, 34, 35
gravity 19, 23, 27
Great Bear see Ursa Major

Halley's Comet 28
Hercules 32, 33, 34, 50
Hyades 35
Hydra 32

infrared ray 52, 55
Io 22, 23

Jansky, Carl 55
Jupiter 12, 22-3, 24, 25, 29, 39, 46, 57, 61

Leo 32, 33, 34
Libra 32
lightwave 54-5
lightyear 46, 47
Little Bear see Ursa Minor
Lynx 32, 34
Lyra 43

M 13 6, 32, 33, 58
Magellanic Clouds 48, 49
magnetic field 16, 18, 22, 25, 41
magnetosphere 18, 22
Mariner 10 14, 15
Mars 12, 20-21, 29, 39
Mercury 12, 14-15, 16, 28, 38, 39
meteor 28-9, 57
meteorite 28-9
Milky Way 12, 33, 35, 47, 48, 58, 61
Mizar 33
Moon 12, 15, 16, 19, 28, 30, 37, 38, 61

nebula 42, 43
Neptune 27, 39
neutron star 44
Newton, Isaac 52
night 30

Ophiuchus 32, 34
Orion 32, 34, 35
Orion Nebula 35, 42

Pegasus 32, 34, 35
Perseus 32, 34, 35, 43
Phobos 20
Pioneer 24, 56
Pisces 34
planets 12-27, 38-9, 44, 46
Pleiades 6, 32, 35

Plough, The 32, 33, 34, 35
Pluto 27, 28, 39, 46
Pointers 33
Pole Star (Polaris) 30, 32, 33, 34, 35
Pollux 32, 34, 35
Procyon 32, 34, 35
protostar 42
pulsar 44, 45

quasar 50, 55

radiation 18, 22, 25, 50
radio astronomy 16, 54-5, 58
radio telescope 50, 55, 58
radio wave 52, 54-5, 58
Rasalhague 32, 34
red dwarf 56
red giant 35, 43
Regulus 32, 34
Rigel 34, 43
Ring Nebula 43

Sagittarius 32
Saturn 1, 24-5, 27, 39, 61
Scorpius 32, 33
seasons 30-31
Serpens 32
Seven Sisters see Pleiades
Sirius 35, 43
Solar System 12-27, 28, 39, 46-7, 56, 58
space telescope 55
spectrum 40, 52, 55
Spica 32, 33
stars 12, 31, 32-5, 38, 39, 40-5, 46, 48-9, 50
Summer Triangle 33, 34, 35
Sun 12-13, 15, 16, 18, 19, 21, 24, 27, 28, 30-31, 36-7, 38, 39, 40-41, 42, 43, 46, 48
sundial 30, 31
sunspot 41
Sun-storm 40, 41
supernova 38, 44

Taurus 32, 34, 35
telescope 52-3, 54, 55, 60-61
Titan 25
Triton 27

UFO 56-7
ultraviolet ray 41, 55
Universe 12-13, 46-7, 50
Uranus 26-7, 39
Ursa Major 33
Ursa Minor 32, 33, 34

Vega 32, 33, 34, 35
Veil Nebula 7
Venus 12, 15, 16-17, 38, 39, 57, 61
Viking 20, 21
Virgo 32, 33, 47, 49
Voyager 1, 22, 25, 26, 27, 56, 57

white dwarf 43, 44
Wolf 359 43

X-ray 41, 52, 54-5

zodiac 39

Acknowledgments

The authors and publisher would like to thank the following for permission to reproduce the photographs in this book: Armagh Planetarium; California Institute of Technology and Carnegie Institution of Washington; Hale Observatories; Richard & Dolores Hill; Robert Little; Drs Foy and Bonneau; Mansell Collection; NASA; National Maritime Museum, Greenwich; New Scientist; Royal Astronomical Society, London; Royal Observatory, Edinburgh; Science Photo Library; Sky Publishing Corporation; Strasenburgh Planetarium, Rochester, N.Y.; University of Arizona; US Naval Observatory, Washington.

The illustrations were prepared by: Chris Forsey; Terry Hadler; Michael Roffe; Studio Briggs; Technical Print.